1978

NEW WORLDS

IN

CHEMISTRY

# NEW WORLDS IN CHEMISTRY

Martin Sherwood

**Second, Revised Edition of *The New Chemistry***



BASIC BOOKS, INC., PUBLISHERS

NEW YORK

Library of Congress Catalog Card Number: 78–54499
ISBN: 0–465–05102–2
Manufactured in the United States of America
10 9 8 7 6 5 4 3 2 1

# Contents

## Illustrations



## Tables

## Acknowledgements

An earlier, longer version of this book (published in the United States of America by Basic Books Inc. as *The New Chemistry*) was read, prior to publication by Mr Martyn Berry of Chislehurst and Sidcup School for Boys, Professor Malcolm Frazer of the University of East Anglia, Dr Jeff Thompson of the University of Oxford, and by my father-in-law, the late Mr A. W. Bennett. In addition, a large part of that text was read by Mr Robert Bruce, also of Chislehurst and Sidcup School for Boys. Miss Lynette Hamblin of the Chemical Society read parts of what is now Chapter 6 and the whole of Chapter 7, and Dr D. G. Jones of ICI Ltd read what is now Chapter 5. For the present text, I have condensed and rewritten the material on chemical theory (Chapter 1 and part of Chapter 2), and the present version of this material has been read by Mr Berry, Dr Thompson, and Dr Tom Faber of Corpus Christi College, Cambridge. In addition, I have made a number of other changes which, though minor, seem to me to improve the material and bring it up to date. For the patience and helpful comments of all those who read the manuscript, or parts of it, I am most grateful, although I should stress that any shortcomings in the final version are my responsibility alone. I am also indebted to Alan Middleton who prepared the illustrations and to Miss Louise Thomas who typed most of the manuscript.

For the use of copyright material, I should like to thank the following: the American Association for the Advancement of Science for material from their journal, *Science*; the American Philosophical Society and Sir Peter Medawar for the excerpt from *Induction and intuition in scientific thought*; Victor Gollancz Ltd and Barry Commoner for the excerpts from the latter's *Science and Survival*; the Nobel Foundation for the excerpt from Sir George Porter's Nobel speech; the Organization for Economic Co-operation and Development for the excerpt from an article by J. A. Campbell, originally published in their *Chemistry today – a guide for teachers*, and for the excerpts from *Science, growth and society – a new perspective*; the Clarendon Press, Oxford, for the excerpts from Melvin Calvin's *Chemical Evolution*; the Royal Institute of Chemistry for excerpts from R. O. Gibson's *The discovery of polythene*; the publishers of

*Scientific American* for excerpts from articles by Harvey Brooks and Raymond Bowers, Bryce Crawford Jr, Von R. Eshleman, Ruth Hubbard and Allen Kropf, and Max Perutz; and George Weidenfeld and Nicolson for the excerpts from J. D. Bernal's *The Origin of Life*.

To the numerous people who, at one time or another, taught me chemistry, I can only say that I hope this book provides some evidence that their time was not wholly wasted. Lastly, but by no means least, my thanks go to my wife, Elizabeth, for her tolerance and understanding.

Martin Sherwood

April 1978

# Introduction

Chemistry is one of the means by which man describes reality. As one of the sciences, it shares the accolade bestowed by the Nobel prizewinning biologist Sir Peter Medawar when he wrote that, 'Science, broadly considered, is incomparably the most successful enterprise human beings have ever engaged upon.'[1] Chemistry is a central science, not only because of its numerous and fundamental connections with other branches of science, but also because of its wide-ranging influence on the way we live.

It can be argued that only a few inventions, such as the transistor, have changed society in a major way; on the other hand, there are literally thousands of inventions which slowly alter the background scenery against which human lives are played out. Although it is often not possible to claim that these inventions have changed society, it can be said that, without them, society would have had to move in different directions. Without synthetic rubbers, the modern automobile industry would not have been able to develop as it has. Without modern fertilizers, much less food would be grown and the percentage of starving people in the world would be much higher (or, alternatively, there would be fewer people). Without modern synthetic drugs, many humans would lead shorter or less healthy lives. These are just three examples of the way in which modern chemistry contributes to the structure of modern society.

Chemistry is, of course, not just a matter of applied science. Developments in chemical research in the last few years have greatly increased human understanding of living processes. Much of this work has been the result of collaboration between chemists and biologists, just as physicists and engineers have called in chemists to help them with such developments as the transistor and electronic microcircuitry.

The science of chemistry can be traced back to early civilizations. The preparation of metals, such as copper, from mineral ores – practised 6,000 years ago – is a primitive form of applied chemistry. The classical Greek view of a world made up from 'atoms' has some parallels with modern atomistic chemistry. Throughout history one can chart the slow growth of

1. P. B. Medawar, *Induction and Intuition in Scientific Thought*, Memoirs of the American Philosophical Society, vol. 75, 1969.

chemistry, sidetracked many times by theories with which we no longer agree, then gathering momentum in about 1800 and proceeding into the twentieth century.

World War II was a watershed in terms of science as an activity. Since the end of that war all developed countries have made gigantic efforts to educate more people as scientists. Until recently, most of those trained as scientists have practised as scientists, so the rapid growth of scientific manpower has been accompanied by an explosive growth in the quantity of scientific data at our disposal. This book is concerned primarily with what has happened in chemistry during the past thirty years.

All sciences are based on theories; because many of these are long lived, it has been necessary to introduce some theoretical ideas that originated before the war. However, although a theory may not be new, at any particular time it has a contemporary interpretation. In the first chapter of this book I have tried to describe the basic theoretical concepts of chemistry in an up-to-date manner. The reader familiar with basic chemistry may wish to do no more than skim through this chapter before moving to the 'reality' of modern chemistry. This begins in Chapter 2 with an account of some of the new tools – many of them based on twentieth-century discoveries in physics – which permit the chemist to obtain the data necessary for his science.

Some of these tools have proved useful in elucidating how the planets of our solar system formed according to chemical laws and how, by following the same laws, the first living beings emerged on Earth from inanimate collections of chemical raw materials (Chapter 3). Documentation of possible processes by which living organisms may have originated has engaged the attention of many groups of scientists since the first experiments in the field in the early 1950s. It is but one example of the fascination of chemists with biological systems. Chapter 4 looks at some of the diverse ways in which chemists have helped to unravel many of the intricacies of life in the present-day world of insects and microbes as well as of man.

Some recent discoveries in biological chemistry, such as the elucidation of the nature of insect sex-attractants and their synthesis in the laboratory, have been reflected in new products from the chemical industry in the past few years. The chemical industry covers an enormous span, which includes the manufacture of valuable materials needed in only small quantities, such as pharmaceuticals, as well as the bulk production of plastics and other polymers. In terms of scale, the industry is dominated by petrochemicals. Based on the utilization of oil as a raw material rather than a source of energy, this sector of the industry has grown enormously in the past thirty years.

As soon as one turns to industry, another, younger science comes into play – economics. Because of the importance of the interactions between chemistry and economics, Chapter 5 is devoted partly to describing the petrochemical industry in a way that highlights these interactions. It must be stressed, however, that the economics does not reach the level of sophistication of some of the chemistry, such as the development of metal-containing catalysts for the production of polymers with clearly defined three-dimensional structures which confer on them much more useful properties then are found in the corresponding random structures.

Molecular engineering – as such conscious design of materials is sometimes called – is gradually spreading throughout industrial chemistry, and industrial chemistry is gradually spreading throughout the whole of civilized life, as Chapter 6 tries to make clear. But that which is found everywhere will be found sometimes in the wrong place – which is one definition of pollution. It would be deceitful to pretend that man's knowledge of chemistry, and the uses he has made of it, have not caused social problems. Some of these are described in the first part of Chapter 7. The remainder of this final chapter is given over to a brief discussion of the role of the scientist as a citizen, and of where chemistry may go in the next few years.

The chemistry included in this book is only a selection of the vast amount of information available. It is, I hope, a representative selection, which will give the reader some picture of what chemistry is like now. One thing it will not do is teach anyone enough about chemistry to enable him to go into a laboratory and make new discoveries. Most of us probably have no real desire to push back the frontiers of scientific knowledge, being primarily concerned with other aspects of the world around us. But it is increasingly important to know how chemistry can be used. Do we want cheaper detergents if it means that fish cannot live in our rivers as a result of these detergents being poured down the sink? Without the background necessary for informed decisions, we cannot decide how chemistry ought to be used which is, as it has always been, a question all of us must try to answer if society is to develop democratically.

# CHAPTER 1
## THE ELEMENTS OF CHEMISTRY

From early childhood, most people take for granted that they live in a real world, surrounded by material objects. The matter around us, and of which we too are composed, has various properties with which we are familiar. It may be hard or soft, dull or shining, for example. It may seem permanent or, like a rotting fish, be seen to undergo changes. For centuries, but particularly in the past hundred years during which they have developed sophisticated instruments and experimental techniques, scientists of various sorts have been exploring the principles which underlie the world of appearance.

Material objects have been catalogued, subdivided and recatalogued. Their constituent parts have been analysed and from these parts new materials have been synthesized. Although the task is one that continues to ever-increasing depths, there are now sophisticated models according to which we can explain the behaviour of matter at different levels.

In the scientific approach to matter, the chemist's role is to describe and understand nature at the molecular level. The physicist, on the other hand, is concerned with the atomic and subatomic building blocks from which all molecules are constructed. It is difficult to draw a clear dividing line between the two approaches; and, in order to understand the recent advances in molecular science with which this book is concerned, it is necessary at the outset to move into the physicist's domain.

The idea that all matter is built from minute atoms is at least as old as the Greek philosophers, Leucippus and Democritus, who flourished between 450 and 400 B.C. However, to them the atom was essentially indivisible – the ultimate building block of matter. That this is definitely not so became apparent in the final decade of the nineteenth century. In 1876, the British scientist Sir William Crookes discovered cathode rays; in 1894 these were given the name electrons. In the following year, it was shown that these electrons carried a negative charge and, shortly thereafter, another British scientist, J. J. Thomson, showed that the electron was about 2,000 times lighter than the smallest known atom.

Thomson showed that the electrons had been torn off atoms by detecting also the heavier fragments left behind. As atoms in their normal state are electrically neutral, those which have lost one or more electrons carry a positive charge. They are known as ions. Once an electron has been torn off one atom, it may attach itself to another, again forming an ion – but in this case, a negatively charged ion.

Little more than sixty years ago, Lord Rutherford showed that at the heart of every atom is a dense nucleus, in which most of the atom's mass and the whole of its positive charge is concentrated. Thus, it was shown that atoms are not indivisible, but consist of at least two types of smaller particle: the electron, and a particle with an equal, but positive electrical charge called the proton.

By this time, chemists had shown that there were many different sorts of atom, each of which had a characteristic mass. Different types of atom were found to have different numbers of electrons and, as atoms ordinarily have no net electrical charge, different numbers of protons in their nuclei. The number of its protons (called the atomic number, $Z$) characterizes any atom as being an atom of a particular element. However, this information is still not sufficient to account for all the known types of atom. For example, the simplest element, hydrogen, is made up of one proton and one electron. Yet there are two rare forms of hydrogen, called deuterium and tritium, which, while showing sufficient properties in common with ordinary hydrogen for us to be sure that they are atoms of this and no other element, are almost exactly twice and three times as heavy, respectively. An explanation for this came in 1932 with the discovery of a third subatomic building block, the neutron. This particle was found to be like the proton in that it forms part of the nucleus of atoms and has a similar mass; however, as its name implies, it is electrically neutral.

Since World War II, high-energy physicists have added greatly to the list of subatomic particles and have also produced evidence that both the proton and neutron may be built from other, more fundamental particles. Recent work in this field has been described by John Taylor in *The New Physics* (197?). However, the chemist is content to stick with the electron, proton and neutron as the fundamental building materials in his branch of science.

He accepts that all atomic nuclei are built of protons and neutrons (with the exception of common hydrogen, which has no neutrons) and that the number of protons ($Z$) determines the chemical species to which an atom belongs. The total number of protons and neutrons in an atom is called the mass number, $A$. Nearly every element in nature consists, like hydrogen, of a mixture of isotopes which share the same value for $Z$, but differ in $A$ because the numbers of neutrons in their nuclei vary. Surrounding the nucleus of each atom are $Z$ electrons to achieve electrical neutrality.

**Transmutation of elements**

There are more than a thousand known isotopes; however, the number of different elements which occur naturally on Earth is only about ninety. This is because large nuclei appear to be inherently unstable. There is an as yet poorly understood force, called the nuclear force, which holds neutrons and protons together. At the same time, the positive electrical charges of protons make them repel one another. As the size of a nucleus increases, the repulsive force outweighs the attractive force, and the nucleus becomes unstable on that account.

Large nuclei, like that of uranium (with ninety-two protons) tend to break up spontaneously into smaller ones. This they can do in a number of ways, all of which are accompanied by a change in the number of protons, which also means that the nucleus changes to that of a different chemical element.

The naturally occurring processes of chemical transmutation are decay processes – large nuclei become smaller. However, about forty years ago Irene Joliot-Curie and her husband Frederic Joliot showed that in the laboratory heavier elements may be built up from lighter ones. They discovered that by bombarding atoms of aluminium ($Z = 13$) with radiation from polonium, an element discovered by Irene's parents, Marie and Pierre Curie, they could produce phosphorus ($Z = 15$). This raised the possibility that quite new elements, heavier even than uranium, which was then believed to be the heaviest naturally occurring terrestrial element, could be created artificially.

During World War II, it was discovered that, by bombarding uranium atoms with neutrons, a new element christened neptunium with ninety-three protons was produced. (After it has hit the uranium nucleus, the neutron undergoes a process called beta decay, by which it turns into a proton and an electron, and the electron then escapes.) Subsequently, by bombarding various heavy elements with increasingly large particles, such as carbon nuclei, it has been possible to produce traces of all the elements with proton numbers from 93 to 106. However, there are still unsettled disputes between American and Soviet scientists as to who first synthesized elements 104, 105 and 106. Until these are settled, the names will not be fixed. More recently (September 1976), the Soviets have also claimed the synthesis of element 107.

An interesting sideline to the work on synthesis of new elements, much of which has been carried out by Glenn Seaborg and his collaborators in the United States, is the possibility that some of the elements beyond uranium may not be really new. For many years, it was believed that uranium was the heaviest naturally occurring element, and that any heavier elements which may once have existed on Earth had long since spontaneously decayed. However, after studying the likely chemistry of the transuranics, Seaborg suggested that a search be undertaken to look for traces of such elements in rocks composed of elements to which they would be related in electronic structural terms. In September 1971, uranium was deposed as the heaviest naturally occurring element when Seaborg announced that an isotope of plutonium, the element with ninety-four protons in its nucleus, had been identified in samples of the mineral bastnasite.

### The atom as a solar system

Seaborg's detective work had, as its basis, the belief that it is possible to group elements according to their electronic structure. The way in which an atom's electrons are distributed about its nucleus is the basic classification of chemistry. From it, as Seaborg's example shows, the chemist derives much of his information about how different elements will behave.

The underlying rationale for this is that it is through its electrons that one atom influences another. When two atoms approach each other, except under highly specialized conditions, their nuclei never come into contact; both atoms retain their elementary identity. However, their electrons may interact readily in different ways – so that, for example, a chemical bond, or a pair of ions (one positive and the other negative) are formed.

Atoms are very small; much too small to be seen by the naked eye, although the best modern electron microscopes come close to 'seeing' individual atoms. A typical atom is about one ten-millionth of a millimetre in diameter – that is, roughly one million times less than the thickness of this sheet of paper. However, the nuclear diameter is 10,000 times less than this. Consequently, nearly all the space which an atom takes up is the domain of its electrons.

The volumes of different atomic nuclei are closely proportional to their mass numbers, $A$. However, there is no simple relationship between atomic volumes and the numbers of extranuclear electrons, $Z$. This implies that, while protons and neutrons pack tightly together in the nucleus, extranuclear electrons are ordered differently. They occupy so much more space than the nucleus that one cannot imagine them being packed together at all. Yet, because they have negative electrical charge, they must be strongly attracted to the positively charged nucleus. Why do they not fall into this dense agglomeration of positive charge?

In another field, astronomy, a similar question was asked, and answered, many years ago: Why do the planets not fall into the Sun as a result of gravitational attraction? The answer to this question is that the centrifugal force of the planets' motion balances out the attractive force of the Sun. Early in the twentieth century, the Danish physicist Niels Bohr proposed that electrons move around their nucleus so that its attractive force is balanced in a similar fashion.

Bohr was trying to explain some spectroscopic data which indicated that hydrogen atoms can exist in different, discrete states, each with a different ionization energy. The ionization energy is a measure of the energy needed to separate an electron completely from its nucleus; in the case of hydrogen, the ionization energies (measured in electron volts, eV) for the different states could be calculated from the formula $13{\cdot}6/n^2$, where $n$ was any of the integers, 1, 2, 3, etc. Bohr managed to derive this formula in a convincing way from a single postulate about the angular momentum which an orbiting electron could be supposed to possess.

There was a snag, however. According to the theory of electromagnetism, a charged particle moving in a circular or elliptical orbit ought to emit radiation continuously. But no radiation is emitted by the hydrogen atom, at least not when it is in its most stable state, when $n = 1$. This difficulty remained until the 1920s, when Erwin Schrödinger and others developed quantum mechanics.

## Electron orbitals

The description of electrons provided by quantum mechanics makes it seem seriously misleading to refer to them – as we still do – as 'particles' at all. The word 'particle' conjures up an image of a discrete object, like a super-miniature marble, for example. But because it is so small, the electron is far harder to pin down than this image implies. Travelling at high speed through a cloud chamber or a photographic emulsion, an electron may leave behind it a well-defined track, just as a particle should; but when confined inside an atom, it spreads out into a continuous cloud of charge that completely envelops the nucleus.

Using quantum mechanical equations, it is possible to calculate with great precision just what the average density of this charge cloud must be in a hydrogen atom, and the results justify many features of Bohr's planetary model. In addition to justifying his postulate concerning the angular momentum of the electron, they show that his orbits correctly describe the average distance of the electron from the nucleus in the different states, and lead to the same formula ($13{\cdot}6/n^2$) for the ionization energies of these states. At the same time, they dispose of the paradox concerning radiation, for a rotating cloud of charge is not to be expected to radiate if it is continuous.

Bohr's planetary model has proved so valuable over a long period of time as an aid to the imagination that both physicists and chemists are reluctant to abandon it completely, and they frequently refer to the electrons in atoms and molecules as moving in *orbitals*. The word implies a trajectory less sharply defined than an orbit, and there is no harm in using it, provided its limitations as a model of reality are borne in mind.

To describe the orbital of an electron about a nucleus, it is necessary to specify more than just the integer, $n$. There are, in fact, four quantum numbers, denoted by the letters $n$, $l$, $m$ and $s$. The first, often called the principal quantum number, describes the average separation of the electron from the nucleus, and hence the effort required to force them apart. The second and third are both related to the angular distribution in space of the electron cloud, dictating the overall shape of the orbital it occupies, which may depart from perfect spherical symmetry about the nucleus. The fourth describes the orientation of what is known as the spin of the electron; this is a property which need not be considered in any depth here, but which is important in modern chemistry as a result of the development of the analytical technique known as electron spin resonance spectroscopy (see p. 53).

For each value of $n$, $l$ may have any value from 0 to $n-1$, while $m$ may be 0 or any number not greater than $+l$ or less than $-l$. The spin quantum number can have one of two values, $+\frac{1}{2}$ or $-\frac{1}{2}$.

From simple arithmetic it can be shown that, when the principal quantum number is 1, then both $l$ and $m$ must be zero. However, as the spin can have one of two values, two different quantum configurations can be associated with this quantum number. Similarly, if $n = 2$, eight different configurations are possible; if $n = 3$, then the number is eighteen, and so on. The importance of these numbers is that they provide a theoretical basis to the major classificatory breakthrough of chemistry: the periodic table of the elements. Before passing on to this, there is one important qualification to be attached to quantum numbers.

This was first expressed by the German physicist Wolfgang Pauli in 1926, on an empirical basis, and subsequently established on a more fundamental basis, and is called the Pauli exclusion principle. It states that no two electrons can occupy an orbital in which all four quantum numbers are the same. Consequently, the numbers derived above for the number of possible quantum configurations for any value of $n$ are the upper limit of electrons that can go into orbitals with that principal quantum number.

## Periodicity of the elements

Just over a century ago, the Russian chemist Dmitri Mendeleyev drew up a table of the elements which arranged them in behaviourally similar groups, and in increasing order of atomic mass. Chemists had been striving for some time to produce a table which would classify the elements in some way, but with limited success. Mendeleyev's table, although it followed its predecessors by listing known elements in order of increasing atomic mass, differed from these by leaving gaps, where the behaviour of certain elements dictated this. The intuitive leap made by Mendeleyev was vindicated in later years as further elements were discovered which fitted neatly into the spaces he had left blank. His original table contained only sixty-three elements; the modern version, as shown in Figure 1, contains many more.

It is possible to justify Mendeleyev's organization completely on quantum mechanical grounds. For example, lithium (Li, $Z = 3$) and sodium (Na, $Z = 11$) are both soft, metallic elements which react violently with water. Because of their behavioural similarities, Mendeleyev placed them in the same group. If we examine these so-called alkali metals in electronic structural terms, it is possible to see why they are related to one another.

The lithium nucleus contains three protons; consequently, a neutral lithium atom has three electrons. Two electrons can have the quantum number $n = 1$, because there is an orbital associated with this quantum number which can hold two electrons. (Note that an orbital describes the shape of the space which an electron occupies; the spin value of an electron has no effect on the shape of an orbital, whereas the other three quantum numbers do. Consequently, any orbital can hold two electrons, provided they have different spin values.) This leaves one electron over. Because of the exclusion principle, it must have a higher value of $n$. As $n$ increases, so does the energy of an electron with that quantum number. Consequently, in the most stable (lowest energy) state of the lithium atom, the principal quantum number for the third electron will be $n = 2$. As energy increases also with increasing values of $l$, this electron will be in the orbital described by $l = 0$ (and, consequently, $m = 0$).

Orbitals in which $l = 0$ are called *s* orbitals. So the electron configuration for the lithium atom in its lowest energy or 'ground' state, can be written in a shorthand form as $1s^2 2s^1$, showing that there are two electrons (indicated by the superscript number) in the *s* orbital with $n = 1$, and one in the *s* orbital with $n = 2$. Because the $n = 1$ level is full and can hold no more electrons it is called a closed shell.

Sodium has eleven electrons, so, in its ground state, it too will have the

## THE PERIODIC TABLE OF THE ELEMENTS

| IA | IIA | IIIB | IVB | VB | VIB | VIIB | VIII | | | IB | IIB | IIIA | IVA | VA | VIA | VIIA | NOBLE GASES |
|---|---|---|---|---|---|---|---|---|---|---|---|---|---|---|---|---|---|
| 1 H | | | | | | | | | | | | | | | | 1 H | 2 He |
| 3 Li | 4 Be | | | | | | | | | | | 5 B | 6 C | 7 N | 8 O | 9 F | 10 Ne |
| 11 Na | 12 Mg | | | | | | | | | | | 13 Al | 14 Si | 15 P | 16 S | 17 Cl | 18 Ar |
| 19 K | 20 Ca | 21 Sc | 22 Ti | 23 V | 24 Cr | 25 Mn | 26 Fe | 27 Co | 28 Ni | 29 Cu | 30 Zn | 31 Ga | 32 Ge | 33 As | 34 Se | 35 Br | 36 Kr |
| 37 Rb | 38 Sr | 39 Y | 40 Zr | 41 Nb | 42 Mo | 43 Tc | 44 Ru | 45 Rh | 46 Pd | 47 Ag | 48 Cd | 49 In | 50 Sn | 51 Sb | 52 Te | 53 I | 54 Xe |
| 55 Cs | 56 Ba | 57 La* | 72 Hf | 73 Ta | 74 W | 75 Re | 76 Os | 77 Ir | 78 Pt | 79 Au | 80 Hg | 81 Tl | 82 Pb | 83 Bi | 84 Po | 85 At | 86 Rn |
| 87 Fr | 88 Ra | 89 Ac† | 104 ‡ | 105 ‡ | | | | | | | | | | | | | |

* *Lanthanide series*

| 58 | 59 | 60 | 61 | 62 | 63 | 64 | 65 | 66 | 67 | 68 | 69 | 70 | 71 |
|---|---|---|---|---|---|---|---|---|---|---|---|---|---|
| Ce | Pr | Nd | Pm | Sm | Eu | Gd | Tb | Dy | Ho | Er | Tm | Yb | Lu |

† *Actinide series*

| 90 | 91 | 92 | 93 | 94 | 95 | 96 | 97 | 98 | 99 | 100 | 101 | 102 | 103 |
|---|---|---|---|---|---|---|---|---|---|---|---|---|---|
| Th | Pa | U | Np | Pu | Am | Cm | Bk | Cf | Es | Fm | Md | No | Lr |

‡ No official names nor symbols have been adopted for these elements.

FIGURE 1 The periodic table of the elements. This version of the table ranks elements with the same number of electron layers in horizontal columns, with the number of electrons in the outermost shell increasing from left to right. Because of the anomalous way in which electron shells are filled among the higher elements, two rows – the lanthanides and actinides – are shown at the bottom of the table. With minor exceptions, elements ranked vertically have the same number of electrons in their outermost shell, and thus show behavioural similarities. Lithium, sodium and potassium, for example, all ionize relatively easily to form positively charged ions. Although similar, elements in vertical columns are not identical in their behaviour; like families, the members have some similarities but also differ from one another in various ways.

$n = 1$ shell closed. The next eight electrons go into the second ($n = 2$) shell, thus filling it completely to make it a closed shell also. Only two of these electrons can go into the $2s$ orbital without violating the exclusion principle. The other six electrons with $n = 2$ go into three spatially different orbitals where $l = 1$ and $m = -1, 0,$ and $+1$. These have a slightly higher

**Table 1** *The elements, their symbols, atomic (proton)numbers and atomic masses*

| *element* | *symb.* | *no.* | *mass* |
|---|---|---|---|
| hydrogen | H | 1 | 1.01 |
| helium | He | 2 | 4·00 |
| lithium | Li | 3 | 6·94 |
| beryllium | Be | 4 | 9·01 |
| boron | B | 5 | 10·81 |
| carbon | C | 6 | 12·01 |
| nitrogen | N | 7 | 14·01 |
| oxygen | O | 8 | 16·00 |
| fluorine | F | 9 | 19·00 |
| neon | Ne | 10 | 20·18 |
| sodium | Na | 11 | 22·99 |
| magnesium | Mg | 12 | 24·31 |
| aluminium | Al | 13 | 26·98 |
| silicon | Si | 14 | 28·09 |
| phosphorus | P | 15 | 30·97 |
| sulphur | S | 16 | 32·06 |
| chlorine | Cl | 17 | 35·45 |
| argon | Ar | 18 | 39·95 |
| potassium | K | 19 | 39·10 |
| calcium | Ca | 20 | 40·08 |
| scandium | Sc | 21 | 44·96 |
| titanium | Ti | 22 | 47·90 |
| vanadium | V | 23 | 50·94 |
| chromium | Cr | 24 | 52·00 |
| manganese | Mn | 25 | 54·94 |
| iron | Fe | 26 | 55·85 |
| cobalt | Co | 27 | 58·93 |
| nickel | Ni | 28 | 58·71 |
| copper | Cu | 29 | 63·55 |
| zinc | Zn | 30 | 65·37 |
| gallium | Ga | 31 | 69·72 |
| germanium | Ge | 32 | 72·59 |
| arsenic | As | 33 | 74·92 |
| selenium | Se | 34 | 78·96 |
| bromine | Br | 35 | 79·90 |
| krypton | Kr | 36 | 83·80 |
| rubidium | Rb | 37 | 85·47 |
| strontium | Sr | 38 | 87·62 |
| yttrium | Y | 39 | 88·91 |
| zirconium | Zr | 40 | 91·22 |
| niobium | Nb | 41 | 92·91 |
| molybdenum | Mo | 42 | 95·94 |
| technetium | Te | 43 | 98·91 |
| ruthenium | Ru | 44 | 101·07 |
| rhodium | Rh | 45 | 102·91 |
| palladium | Pd | 46 | 106·40 |
| silver | Ag | 47 | 107·87 |
| cadmium | Cd | 48 | 112·40 |
| indium | In | 49 | 114·82 |
| tin | Sn | 50 | 118·69 |
| antimony | Sb | 51 | 121·75 |
| tellurium | Te | 52 | 127·60 |
| iodine | I | 53 | 126·90 |
| xenon | Xe | 54 | 131·30 |
| caesium | Cs | 55 | 132·91 |
| barium | Ba | 56 | 137·34 |
| lanthanum | La | 57 | 138·91 |
| cerium | Ce | 58 | 140·12 |
| praseody-mium | Pr | 59 | 140·91 |
| neodymium | Nd | 60 | 144·24 |
| promethium | Pm | 61 | — |
| samarium | Sm | 62 | 150·40 |
| europium | Eu | 63 | 151·96 |
| gadolinium | Gd | 64 | 157·25 |
| terbium | Tb | 65 | 158·93 |
| dysprosium | Dy | 66 | 162·50 |
| holmium | Ho | 67 | 164·93 |
| erbium | Er | 68 | 167·26 |
| thulium | Tm | 69 | 168·93 |
| ytterbium | Yb | 70 | 173·04 |
| lutetium | Lu | 71 | 174·97 |
| hafnium | Hf | 72 | 178·49 |
| tantalum | Ta | 73 | 180·95 |
| tungsten | W | 74 | 183·85 |
| rhenium | Re | 75 | 186·20 |
| osmium | Os | 76 | 190·20 |
| iridium | Ir | 77 | 192·22 |
| platinum | Pt | 78 | 195·09 |
| gold | Au | 79 | 196·97 |
| mercury | Hg | 80 | 200·59 |
| thallium | Tl | 81 | 204·37 |
| lead | Pb | 82 | 207·20 |
| bismuth | Bi | 83 | 208·98 |
| polonium | Po | 84 | — |
| astatine | At | 85 | — |
| radon | Rn | 86 | — |
| francium | Fr | 87 | — |
| radium | Ra | 88 | 226·03 |
| actinium | Ac | 89 | — |
| thorium | Th | 90 | 232·04 |
| protac-tinium | Pa | 91 | 231·04 |
| uranium | U | 92 | 238·03 |
| neptunium | Np | 93 | 237·05 |
| plutonium | Pu | 94 | — |
| americium | Am | 95 | — |
| curium | Cm | 96 | — |
| berkelium | Bk | 97 | — |
| californium | Cf | 98 | — |
| einsteinium | Es | 99 | — |
| fermium | Fm | 100 | — |
| men-delevium | Md | 101 | — |
| nobelium | No | 102 | — |
| lawrencium | Lr | 103 | — |



Note: Elements 104 to 106 are not included as their syntheses are still disputed (see p. 19). Atomic masses – sometimes loosely called atomic weights – are worked out relative to the mass of the most common isotope of carbon ($^{12}C$); in this table, masses are shown to no more than two decimal places.

energy than the *s* orbitals, and are called *p* orbitals. There is still one electron unaccounted for, and this must go into the lowest energy orbital where $n = 3$. Again, this is an *s* orbital, and the total electron configuration for sodium can be expressed as $1s^2 2s^2 p^6 3s^1$. When this is compared with lithium, one similarity is immediately apparent: in their outermost shells, both elements have a single electron in the same type of orbital. The outermost shell is the one which will come into contact first with any other atom – and, in consequence, the similar electron configuration of the outermost shells of these two elements accounts for the similarities in their behaviour.

Below sodium in the periodic table is the element potassium (K, $Z = 19$). Following the reasoning used above, one could assign the first eighteen of these electrons to the orbitals $1s^2 2s^2 p^6 3s^2 p^6$. Following the quantum numbers procedure, the next orbital to be filled should be one in which $n = 3$ and $l = 2$ (where $l = 2$, we have a so-called *d* orbital; where $l = 3$, an *f* orbital). However, as the number of electrons about a nucleus increases, the tightness with which they are bound – and, consequently, their energy level – depends on the shape of the orbital as well as on the principal quantum number *n*. Because of this, the eight electrons for which $n = 3$ and $l = 0$ or 1 are sufficient to form a closed shell by themselves. The nineteenth electron of potassium goes into a 4*s* orbital, giving potassium the configuration $1s^2 2s^2 p^6 3s^2 p^6 4s^1$: hence its similarity to lithium and sodium.

The irregularities in the filling of orbitals, which are shown graphically in Figure 2, explain the shape of the periodic table (Figure 1). For example, the 3*d* orbitals are of lower energy than the 4*p* orbitals. Consequently, after the 4*s* orbital is filled (calcium, Ca, $Z = 20$), the 3*d* orbitals then fill. After they are filled (zinc, Zn, $Z = 30$), the 4*p* orbitals begin to fill. The situation becomes even more complicated after lanthanum. This element has the first three shells closed, and then the configuration $4s^2 p^6 d^{10} 5s^2 p^6 d^1 6s^2$. After this, the 4*f* orbitals (which can hold fourteen electrons) begin to fill up. So that the table does not become impossibly wide, the series of elements in which this is happening – called the lanthanide series – is set apart from the rest of the table, together with the actinide series, where a similar progression takes place in the next higher set of shells.

To return to the alkali metals, another property which they share in common is a low first ionization energy. In each case, the single outermost electron is rather effectively shielded from the nucleus by completed shell(s) of electrons. Consequently, it is fairly easy to remove, requiring only about 5 eV. As Figure 3 shows, when we move across the periodic

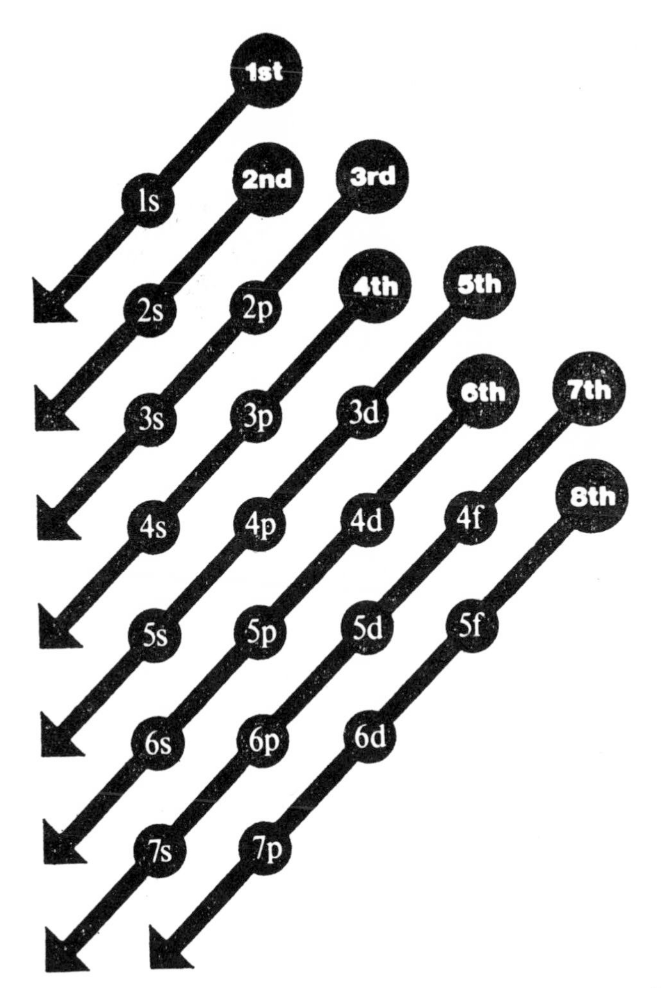


FIGURE 2 The order in which electron orbitals are filled as atoms build up. The progression is straightforward at first, then becomes more complex as the total number of electrons increases; the irregularities are caused by the interactions of electrons with one another. Even this figure is a simplification; for example, one electron appears in the *5d* level before any electrons go into the *4f* level, but no further electrons enter *5d* until *4f* is filled.

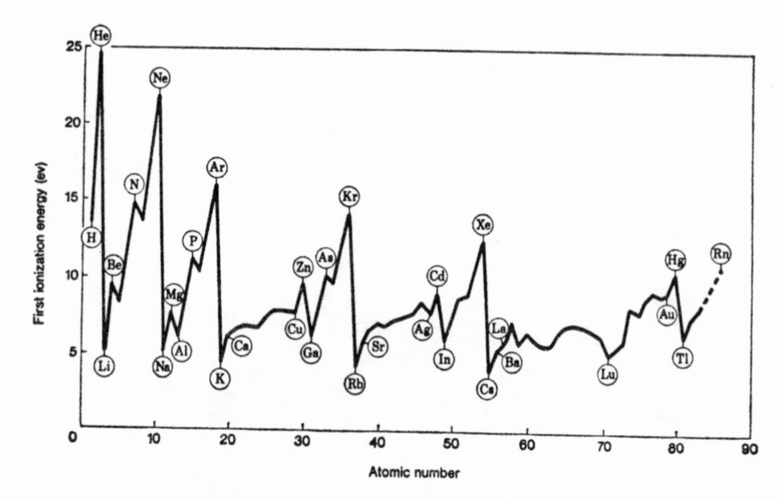


FIGURE 3 Ionization energies of the elements. This graph indicates the amounts of energy required to remove one electron from an atom. Note that elements which occur immediately prior to those in which a new layer starts to fill up occur as a series of peaks (helium, neon, argon, krypton and xenon). They require more energy to release one of their electrons than do their neighbours. Each dot in the graph represents an element, although only a few of the elemental symbols are shown (the others may be worked out from the atomic numbers given on the horizontal axis).

table from left to right at any level, the ionization energy increases, reaching a maximum when a shell is closed, then falling again.

## Bonds between atoms

We have seen that it is possible to remove electrons from an atom to form positively charged ions. From the quantum rules, we know also that most atoms have unfilled electron shells: that is, in theory, their outermost electron shells could contain more electrons than they actually do. In the case of fluorine (F, $Z = 9$), for example, an additional electron would complete the $n = 2$ shell. Completed shells, as the ionization energies of helium, neon, argon, krypton and xenon indicate, are particularly stable configurations. In fact, these five elements were at one time called the inert gases, because it was believed they were so stable that they would not react with any other atoms. However, this has now been disproved (see p. 35).

It seems reasonable that where an element has a nearly closed shell, such as fluorine, it might have an affinity for additional electrons. This is the case: the affinity of fluorine towards an additional electron is about 3·4 eV. To put this another way, 3·4 eV are needed to remove an electron from the negatively charged fluorine ion, $F^-$. This is only slightly less than the energy required to remove an electron from a neutral atom of one of the alkali metals. Consequently, if a neutral fluorine atom and a neutral lithium atom came close together, one might expect some competition between them for the lithium atom's outermost electron.

Many other atoms also have electron affinities which can be expressed in quantitative terms; however, fluorine has one of the highest values. The simplest atom, hydrogen, has an electron affinity about four and a half times smaller than fluorine. However, this still means that a hydrogen atom exerts an attractive force on other electrons.

As a free electron approaches a neutral hydrogen atom, the energy of the atom is lowered. But the electron does not have to be free; it may belong to a second hydrogen nucleus. Then, if the two neutral hydrogen atoms approach sufficiently close together, each nucleus will attract the other's electron. If the two atoms meet under favourable conditions, this may result in the two nuclei sharing the two electrons. In this case, a hydrogen molecule ($H_2$) is formed, in which the two atoms are held together by a chemical bond, which takes the form of a new orbital, called a molecular orbital, which spreads around both nuclei and contains both electrons.

The shape of this orbital can be calculated; so can the distance between the nuclei – called the bond length – for it is part of the essence of the bond that the two nuclei held together by the shared electrons remain nearly the same distance from each other as long as the molecule exists.

Bond lengths can be calculated from mathematically produced curves like that in Figure 4. This shows a 'potential well' for the hydrogen molecule. Along the horizontal axis we measure the distance between the two nuclei. As they come closer together, electron sharing begins and the graph dips down the vertical energy scale. After a while the nuclei get so close that, if they were pushed any closer together, the repulsion of the two positively charged nuclei would be stronger than the cementing effect of the electrons. Consequently, the system needs energy for this to happen, and the curve swings rapidly upwards. The very bottom of the curve shows the energy level which corresponds to the bottom of the well and, at the same time, the distance apart of the two nuclei which is the optimum bond length.

Where two identical atoms come together to form a molecule, both will have the same electron affinity and, in consequence, will share electrons equally. However, an atom such as fluorine has a much stronger electron

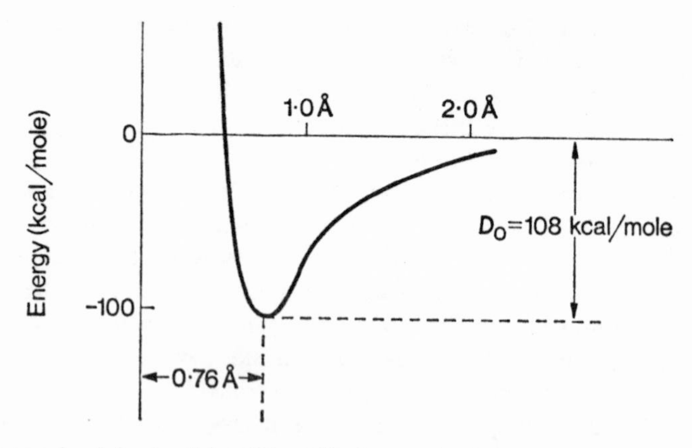


FIGURE 4 A 'potential well' profile for two hydrogen atoms as they approach one another. As they come together, the well gets deeper, indicating that proximity is favoured. However, internuclear repulsion eventually overcomes the attractive force pulling the two atoms together. To make them go even closer, energy must be supplied, and the side of the well shoots up steeply. From a plot such as this, the bond strength and bond length of a hydrogen molecule can be calculated, as can the energy required to stretch the bond by a specified amount.

affinity than does an atom of hydrogen. If atoms of these two elements come together, they can share electrons to form a bond but the fluorine atom gets a much larger share than the hydrogen atom. In terms of the molecular orbital for hydrogen fluoride (HF), there is a much higher electron density around the fluorine atom than around the hydrogen atom. Under certain conditions, notably in solids or solutions, there is a complete take-over of one or more electrons by one species from another. Thus, lithium fluoride, in solution consists of positive lithium ions and negative fluoride ions.

### More complicated bonds

A simple chemical bond, such as that in a hydrogen or hydrogen fluoride molecule involves two electrons. However, more than one electron from an atom may be involved in bond formation. For example, a nitrogen atom needs three electrons to close its outermost shell. When two nitrogen atoms join together to form a molecule of nitrogen ($N_2$), each contributes three of its electrons to the pool of shared electrons. These are arranged in three pairs to form three bonds. Because the electrons have different quantum numbers (although the principal quantum number is the same for them all), they form different types of bond, with differently shaped molecular orbitals. The types of orbital in a particular molecule play an important part in determining its shape; Figure 5 shows the shapes of some simple molecular orbitals.

In general, an atom of an element can form as many bonds as the number of electrons it needs to complete a shell, provided that the shell is already at least half full. But it does not have to form all its bonds with one other atom. For example, a nitrogen atom might have two single bonds to two separate hydrogen atoms, with a third single bond to an oxygen atom. As oxygen readily forms two bonds (it has six electrons in its outermost shell), the oxygen atom may have another bond to a carbon atom which, in turn, may be joined to three other atoms by single bonds. In this way, extremely complex molecules, containing thousands of atoms, can be built up.

So far, we have considered that any bond involves the sharing of two electrons (equally, or unequally, according to the different electron affinities of the atoms involved) between two nuclei, with each nucleus contributing one electron to the bond. Another possibility is that both electrons in a chemical bond may have come from only one of the two atoms involved. Some molecules, even though stable by themselves, may have spare electrons to form further bonds.

In the case of ammonia ($NH_3$), for example, a nitrogen atom is linked to three hydrogen atoms by single bonds, and consequently has shares in six electrons. In addition, there are two more electrons in its outermost shell that are not involved in molecular orbitals. These also can take part in bond formation. Ammonia will bond with the very reactive compound boron trihydride ($BH_3$) to form a complex $H_3N.BH_3$, in which both electrons in the boron-nitrogen bond have been donated by the nitrogen atom, which is consequently called an electron donor. In this way the boron atom has shares in eight electrons altogether, the number it needs to close its second shell.

Boron hydride is so reactive that if there is no electron donor available to form a molecular complex, two molecules of boron hydride link together to form a larger molecule, diborane ($B_2H_6$). There is a superficially similar looking and very common carbon-containing compound called ethane ($C_2H_6$), in which two carbon atoms are singly bonded to each other, and each is bonded separately to three hydrogen atoms. This is a logical arrangement: each carbon atom has four electrons in its outermost shell available for bonding and, in ethane, each carbon atom is linked to other atoms by four single bonds. However, boron has only three bonding

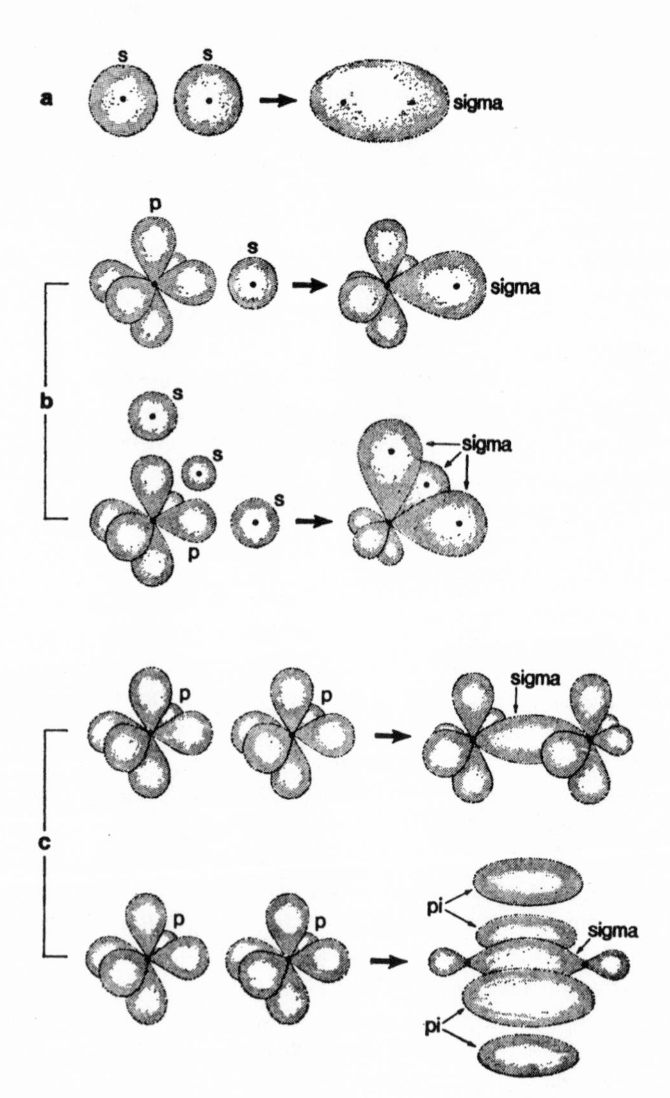


FIGURE 5 The shapes of some simple orbitals: (a) *s* orbitals are spherically symmetrical about a nucleus. When two *s* orbitals on different atoms interact, a sigma bond is formed. (b) *p* orbitals are directional (the different directions are designated *x*, *y* and *z*) and a single *p* orbital from one atom can interact with the *s* orbital from another atom to form a sigma bond; similarly, the three different directional *p* orbitals can each form a sigma bond with *s* electrons from three other atoms. (c) Two *p* orbitals from different atoms can interact to form a sigma bond, but once this has happened, the other-directional *p* orbitals of the two atoms cannot overlap in the same way. If they are to form bonds, they must be directional, unlike sigma bonds, as a result of which they confer rigidity on the resultant molecule, as one nucleus can no longer rotate with respect to the other nucleus without breaking a bond. This leads to 'geometrical isomerism' in doubly bonded molecules (see p. 47).

electrons, so one $BH_3$ cannot meet another $BH_3$ and form a boron-boron bond, for neither has electrons to spare.

When modern techniques of physical chemistry, which allow very precise measurements to be made of the sizes and shapes of molecules, were applied to diborane, it was found that two of the hydrogen atoms bridge the gap between the two boron atoms, so that the molecule has the structure:

```
H\   /H\   /H
  B     B
H/   \H/   \H
```

This, however, does not make sense on the basis of bonding as it has been described so far. We have assumed that a bond is formed when two electrons form a molecular orbital about two nuclei. However, this is an oversimplification. For example, when a molecule contains more than two nuclei, although we may be able to assign two-electron bonds to form neat linkages, in reality the different molecular orbitals may interact slightly with one another. In the case of the B—H—B links in diborane (called 'bridges'), we have a more extreme deviation from the simple 'two electrons equal one bond' model. Each of these two linkages is held together by a three-centre molecular orbital, encompassing all three atoms, but containing only two electrons. The binding between boron and hydrogen atoms in this three-centre bond is weaker than in ordinary boron-hydrogen bonds, like those which hold the other four hydrogen atoms to the boron atoms; none the less, it is a real chemical bond.

One other point about the B—H—B linkage is that it has an identical atom at each end. Each boron contributes one electron to one of the two bridges and, as the atoms at the ends are the same, their electrons have exactly the same energy. It is thus not possible to say which electron in a particular linkage comes from which boron atom. Because of the energy differences between electrons in different atoms, the so-called electron-deficient compounds held together by such bridges are rare, existing (so far as we know) only when holding together atoms of the same element.

This does not mean that they cannot exist in more complex forms. During the 1960s, a host of boron hydrides was discovered in which more and more boron atoms were held together by bonds containing fewer and fewer total electrons. One of the largest of these, discovered in 1963, is $B_{20}H_{16}$. The bigger boron hydrides are held together by a mixture of boron-boron, boron-hydrogen, and boron-hydrogen-boron bonds, with many of the boron-boron bonds being partial bonds held together by 'delocalized' orbitals, covering more than two atoms.

While the story of boron and how it makes do with a shortage of electrons is fascinating, how important is electron-deficiency bonding in general? The answer is that every known metal is made up of atoms of the sort that are ripe for electron-deficient bonding with delocalized orbitals. Like the boron atoms in the large boron hydrides, which organize themselves into tight structures so that their shared molecular orbitals cover as small an amount of space as possible, metals form tightly packed structures, producing regular crystals (an important property of metals) and mobility of electrons (another important property of metals, usually called conductivity). The story of the boron hydrides is also important in broadening our model of chemical bonding to include the idea of delocalized orbitals, which cover more than two atoms. Such delocalized orbitals may occur in molecules even where localized orbitals can be assigned to all the bonds, and may consequently affect the molecule's behaviour. It is even possible that life on Earth may have owed its origin to delocalized orbitals (see p. 80).

### Electrons to excess

Having seen that compounds can form with 'too few' electrons, the obvious question is: Are there any compounds with excess electrons? The answer is yes. These involve another type of orbital, called the non-bonding orbital. Such orbitals involve only atoms which are not adjacent to one another in a molecule, and consequently they do not affect the strengths of chemical bonds, because the separation of the atoms with such an orbital means that there is little interaction between them. However, the reality of such orbitals is important in explaining the existence of some compounds which, according to simpler bonding models, should not exist.

The earliest known example is the tri-iodide ion ($I_3^-$). Iodine, which has an outer electron layer similar to chlorine and fluorine, forms a diatomic molecule ($I_2$) in which two atoms are joined by a single bond, with no residual unfilled orbitals. Similarly, if an iodine atom gains an electron, it has a filled atomic orbital. However, from molecular orbital theory, it is possible to postulate a three-centre iodine molecular orbital, in which three iodine atoms are held together by two electrons, as in the B—H—B bond. But there are still two electrons unaccounted for: these go into a non-bonding orbital which is localized about the two ends of the I—I—I linkage and, therefore, they neither help nor hinder bonding. Consequently, $I_3^-$ has about the same bond strength as $I_2$.

Although many scientific discoveries originate by happy accidents, there are also cases where discoveries have been delayed because they appeared impossible. Electron-excess compounds are a case in point. The tri-iodide ion's existence remained a puzzle until molecular orbital theory explained it logically. Once it had been shown how electron-excess compounds can occur, a host of them was discovered, many quite similar to the tri-iodide ion.

A set of electron-excess compounds, the discovery of which surprised many chemists, is the noble gas compounds. The noble gases were formerly called the inert gases. Generations of chemists were taught that, because they had a complete outer layer of electrons, these elements – helium, neon, argon, krypton, xenon – would not form compounds. But, early in 1962, Neil Bartlett, an English chemist working at the University of British Columbia, mixed together platinum hexafluoride and xenon and found that they immediately reacted to form a compound. This result should not have been totally unexpected. Nearly thirty years earlier, Nobel prizewinner Linus Pauling had predicted on theoretical grounds that some of the noble gases might form compounds. However, attempts to substantiate his hypothesis by experiment had failed, and such experimentation ceased in the belief that inert gas compounds were impossible.

During World War II, work on the atomic bomb, which contained uranium purified by a process that involved formation of uranium hexafluoride, led to much freer availability of fluorine, and development of expertise in its handling. The latter is a tricky business, for fluorine is an extremely unpleasant element which can damage lungs, eyes and skin on contact.

The interest in uranium hexafluoride, which could be vaporized easily – the key to its role in uranium purification – led to increased interest in the fluorides of other metals. A new way to purify a metal is always of interest because of possible commercial application and, after the war, work was carried out at the Argonne National Laboratory in the United States on the preparation of fluorides of other rare metals, such as platinum, osmium, and iridium.

Bartlett was studying the hexafluorides of platinum, ruthenium, and rhenium. He discovered that platinum hexafluoride reacted with oxygen to form an ionic compound. Since during ionic compound formation an electron is transferred from one atom to another, it is reasonable to assume that there may be a relationship between the ease of compound formation and the ease of ionization of a molecule (ionization being the complete removal of an electron). Bartlett noted that the ionization energy of the oxygen molecule was very similar to that of the xenon atom. It was this that led to his crucial experiment.

Although he succeeded in making xenon compounds, Bartlett could not get a reaction between the hexafluoride and krypton, nor did he know what the xenon compounds were. He knew that they were compounds of xenon, but the details of how many bonds were involved and the structures of the compounds were still a mystery.

Bartlett's work was soon followed up at the Argonne Laboratory by Howard Claasen, Henry Selig and John Malm. On 2 August 1962, they mixed together xenon and fluorine and heated them at 400 °C for one hour; the result was xenon tetrafluoride ($XeF_4$). They later found that it was also possible to make two other fluorides of xenon, one containing two atoms of fluorine for each atom of xenon ($XeF_2$), the other containing six ($XeF_6$).

Following this exciting work, many chemists became interested in the noble gas compounds. In the past decade, krypton fluorides have been isolated, as well as a compound of xenon with another halogen element, xenon dichloride ($XeCl_2$). In addition, more complicated compounds, in which xenon was bonded to oxygen were prepared during the 1960s and, in 1974, the first compound involving a xenon-nitrogen bond was reported. The bonding in the simple xenon dihalides is similar to that in the triodide ion (this is not surprising, for $I^-$ has the same electronic structure as Xe), and the bond energies are about the same as those found in the diatomic halogen molecules, such as $I_2$.

Compounds involving other noble gases, or other halogens, have not yet been discovered. This is a function of the size of the atoms involved. The smaller the halogen, the fewer electron shells it has, and the tighter it binds electrons. So it is not surprising that fluorine forms most of the noble gas compounds. Using the same reasoning, the noble gases which most readily form compounds are the heavier ones, with multiple electron shells shielding the nucleus, so that the outermost electrons are relatively weakly held and can easily be incorporated into molecular orbitals. The electrons in the smaller noble gases appear to be too tightly held by their nuclei to permit sharing.

Once the basic idea of a three-centre bond, such as that found in $XeF_2$, is accepted, there is no reason why another neutral molecule of fluorine should not be added to form $XeF_4$, and then another to form $XeF_6$. Why not go even further – what about $XeF_{28}$ or $XeF_{102}$? What must not be forgotten is that, as more atoms are added around a central atom, the environment gradually becomes crowded. Repulsive forces between the electron clouds of adjacent atoms come into play, altering the energy balance until the energy gained by bond formation is outweighed by the repulsion energies, and the result is no more bonds. Consequently, there are no molecules of $XeF_{28}$ or $XeF_{102}$.

This restriction raises an interesting point, which has already been touched upon. If there are repulsive forces between electron clouds, how well can a knowledge of molecular orbital theory, coupled with this concept of repulsion, be used to calculate theoretically the shapes of molecules? Quite well – if the molecules are simple. However, for complex molecules containing more than a few atoms, the computations involved are too complicated for this approach to be reliable. For this reason, nearly all present-day knowledge of molecular shapes comes from use of one of the many tools that chemists and physicists have developed in the past few decades. These methods, and some of the more complex aspects of molecular shapes, are discussed in later chapters, notably Chapter 2.

# *CHAPTER 2*

## THE MECHANICS OF CHEMISTRY

The importance of molecules to the chemist is that they can react with each other, or with various forms of energy, such as heat and light, to produce new and different molecules. In other words, the spatial relationships between atoms change; some chemical bonds break and new ones form. In a simple case, molecules of a single compound, such as the gas chlorine trifluoride ($ClF_3$) may break up to form a mixture of chlorine monofluoride ($ClF$) and fluorine ($F_2$) gases. Alternatively, two different sorts of molecules may come together to produce a single, different product (hydrogen and oxygen molecules react to form water) or a mixture of different products (ethane burns in oxygen to form carbon dioxide and water).

The reactivity of molecules does not seem strange because we are used to seeing chemical reactions in daily life – for example, liquid petrol goes into cars, and a mixture of different vapours emerges from the exhaust system. But are we justified in taking chemical reactivity for granted? In the preceding chapter, we saw that, because of their electron affinities, two hydrogen atoms will bond together to form a hydrogen molecule. Figure 4 (p. 30) indicated that such an arrangement lowers the overall energy of the system – that is, makes it more stable. Similarly, when two hydrogen molecules and an oxygen molecule react to produce two molecules of water, energy is released.

Why then, has all the hydrogen and oxygen in the universe not combined to form water? One immediate answer is lack of proximity. Outer space contains numerous isolated clouds of hydrogen gas; the Earth's atmosphere is 20 per cent oxygen gas. An extra-terrestrial hydrogen cloud, millions of miles from the Earth obviously cannot react with atmospheric oxygen. A chemical bond is very short, and it is reasonable to suppose that if such a bond is to form between two atoms, they must approach each other to within a bond length. In general, we consider it necessary for two molecules to collide before a reaction can occur between them. The number of collisions – and we might expect the rate of a reaction, that is, the speed with which reactants are converted into products, to be connected with the frequency of collisions of reactable molecules – will depend on the state of the reacting substances. Gas molecules, for example, move more freely than molecules in liquid, although, by comparison with solids, the motion of liquid molecules is quite free. However, gases can be kept under low pressure, with few molecules occupying a large amount of space; this will lower collision frequency. On the other hand, under high pressure, collisions should take place more often. In that case, and taking into account the extra stability introduced into the system by reaction, why is it possible to keep a mixture of oxygen and hydrogen gases together in a container at high pressure without any apparent reaction? Like a rock balanced at the top of a hill, why doesn't the reaction roll down into the energy valley and reach greater stability? Why, in the millions of years since the universe began, haven't all the energy-releasing reactions between adjacent molecules taken place to produce a stable, unreactive universe?

The answer is that rocks do not always roll downhill. Often they need a push to overcome inertia: in order to release the rock's potential energy, a small amount of energy must be put into it to get it going. Chemical reactions are the same; they need a push to get them going.

Disregarding for a moment the type of push needed, do all reactions represent a downhill roll? Are there none that are like climbing uphill? A moment's thought shows that there must be some, for many reactions can be seen to work in two directions. We may see a set of reactants being converted into products, while elsewhere those products are being converted back into reactants. J. A. Campbell, professor of chemistry at Harvey Mudd College, California, summed this up when he wrote: 'The iron sides of a blast furnace rust, while inside the furnace rust is changing into iron. . . . At one spot in the forest a tree rots in the presence of oxygen to form carbon dioxide and water, while in another spot the water and carbon dioxide just released by the rotting are reacting to form new wood and oxygen.'[1]

## The energies of molecules

From elementary physics, we know that a boulder perched atop a gulley has potential energy. When it starts rolling, this energy is converted into kinetic or motional energy. Because of the frictional forces between the rock and the ground over which it is rolling, this kinetic energy is converted into heat energy. From the viewpoint of the present investigation, we are interested in the energy of molecules, which chemists call 'free energy'.

When a reaction occurs, there is a change in the free energy of the system. Scientists divide systems into three types. First is the isolated

1. J. A. Campbell, 'Why Chemical Reactions Occur', in *Modern Chemistry*, ed. J. G. Stark, Penguin Books, Baltimore, 1970, pp. 113–140.

system which is never influenced from outside in any way; neither energy nor matter can be introduced into or escape from an isolated system. The only example of such a system is the entire universe. In a closed system, on the other hand, no matter is lost to the outside, nor is any introduced, but the system may exchange energy with its surroundings. A chemical reaction taking place in a sealed container is a closed system. Finally, in an open system, both matter and energy exchanges with the outside can take place, as when a kettle of water is boiled: energy is introduced from the outside and, if we let the kettle boil for a few minutes, water vapour is passed out of the system. There is an exchange of both matter and energy.

Some of the energy change in a reaction may be visible. For example, cellulose – a major component of paper – will burn in air to form carbon dioxide and water. We can detect energy given off by this reaction in the forms of heat and light. If we measure energy changes of this kind, however, we are measuring only a part of the free energy change, called enthalpy. There is a second part, entropy, which, in very simple terms, can be called a measure of the degree of organization of the system.

The free energy change in a reacting system can be calculated from an equation which involves both enthalpy and entropy changes. For a reaction to occur spontaneously, the free energy change in a system at constant pressure and temperature must be negative (ie less than zero). As the equation for calculating free energy change contains a variable (temperature), the spontaneity of a reaction depends to some extent on the temperature at which it takes place. For example, at room temperature, calcium oxide (quicklime) spontaneously absorbs carbon dioxide gas to form calcium carbonate, the main constituent of both limestone and marble. But at temperatures of more than 800 °C, the free energy change is negative for the reverse reaction, which is therefore spontaneous at this temperature. Calcium oxide is made commercially by roasting carbonate rock.

From the size of the free energy change of a system under any given conditions, the extent of the reaction – a measure of its completeness – can be calculated. If two compounds which react together rapidly to form a third single compound are mixed together, then, after a while, if the temperature and pressure are kept constant, we shall probably find that we have a mixture of all three substances (starting materials and product) which remains constant in its proportions: a state of equilibrium has been reached. Of course, if the product is removed from the system – if an open system is used – this equilibrium will not be established. One method used by chemists to maximize the product from a reaction is to remove it as it is formed.

For many reactions, it is possible to consult tables of data which have been built up over the years and, on the basis of these, calculate whether or not the reaction will occur spontaneously and, if so, to what extent. For a flask of hydrogen and oxygen at room temperature and atmospheric pressure, for example, the equilibrium favours spontaneous change of nearly all of it into water (assuming that the proportions of hydrogen and oxygen are 2:1 as they are in the product). What we do not know is how fast the reaction will occur.

## Geometry of reactions

Not only are molecules dynamic, in constant motion; a reaction is also dynamic. It is a process which follows a specific pathway and takes a definite amount of time. Complex molecules may follow more than one pathway to produce different sets of products, but for the moment we will keep a simple reaction model. Because the time taken for an individual molecule to react is usually too small to be measured it does not mean that reactions occur instantaneously. A chemist in a laboratory will be dealing with many millions of molecules, so that the time taken for all, or even a noticeable amount, of them to react can be measured. The study of the rate at which reactions occur is called reaction kinetics, while the examination of the pathways taken is the chemistry of reaction mechanisms.

A key concept in reaction mechanisms is the transition state, or activated complex. Earlier it was stated that it is necessary for two molecules to collide for a reaction to take place; but necessary does not mean sufficient. For example, as molecules occupy space and have definite three-dimensional shapes, it is possible for them to collide so that the parts between which new bonds might form are still too far apart for this to happen. The reaction geometry must be right, or the two molecules will just move apart again.

Assuming that the geometry is right, what happens then? In diagramming the act, we can just erase any existing bonds and draw in the new ones, but in reality there must be a breaking and making of bonds over a period of time. The activated complex is the molecular state in which the alteration of structure takes place. These complexes are rarely isolated, for they usually break up too quickly to be isolated. Because they are more ordered (less entropic) than either starting materials or products in a reaction, such complexes are more energetic. Consequently, energy must be supplied for their formation.

The energy of any collection of molecules is an average of the energies of all the individual molecules within that collection. When two molecules collide, they may exchange small amounts of energy. For example, if we refer back to Figure 4 (p. 30), which relates the internuclear distance and energy of two hydrogen atoms, we can see that, to move the two nuclei apart by 0·1 Å ($10^{-11}$ m) from their most stable distance (0·76 Å) requires about one-thirtieth of the bond strength. Consequently, in a collision, a hydrogen molecule might absorb energy from another molecule by stretching the bond between the two hydrogen nuclei.

Individual molecules can absorb and emit small quantities of energy by the stretching and contraction (vibration) of their bonds. The more bonds there are in a molecule, the greater the number of possible vibrations. Another way in which energy can be absorbed or desorbed is through rotation. A hydrogen molecule can be imagined as being like a minute dumb-bell, the hydrogen nuclei forming the two ends, and the molecular orbital representing the bar that joins them (H—H). Just as a dumb-bell can be rotated about the middle of the joining bar, so a hydrogen molecule can rotate about the centre of its single bond. The speed of this rotation, and consequently the energy of the molecule, can be affected by collisions.

In any collection of molecules, a few will probably have enough energy to form an activated complex. The proportion of such molecules can be calculated; they are the ones that already have the push in them to get over the energy barrier and roll downhill to form products. The height of the barrier is called the activation energy.

When a pair of molecules forms the complex and rolls downhill to form products, energy is released. This may be taken up by unreacted molecules, thus supplying them with the energy required to surmount the activation barrier. A piece of paper does not catch fire spontaneously in air. If you supply energy to a small part of it by focusing the sun's rays on to it through a powerful magnifying glass, once the paper has caught fire it will continue to burn because the initial reaction produces enough energy to get adjacent molecules over the activation energy barrier. Drop a lighted match into a flask of oxygen and hydrogen and the same thing will happen, but more quickly. It explodes.

## Helping molecules to react

Different reactions have high or low activation energies, depending on the reaction pathway. The chemist has a number of ways in which he can supply energy to a collection of molecules to help them over the activation energy barrier – in other words, to get the reaction going at a speed which suits him. The most obvious is to heat the mixture.

An alternative is to raise molecules to an energy level higher than the activation barrier. This can sometimes be done with light; a molecule may absorb light of a specific wavelength and, as a result, have one of its electrons promoted into a different, more energetic orbital (excitation). Overall, this molecule may now be raised about the level of the activation barrier for the reaction. Assuming it makes a satisfactory collision with the molecule with which it is to react, it will probably form the intermediate complex. Light may also be used to induce those reactions in which a single molecule reacts by breaking up, collision with a 'photon' of light energy being sufficient to break some bonds.

As activation energy also depends on reaction pathway, it is possible to effect some difficult reactions (those with high activation energies) by providing an alternative pathway which has a lower activation energy. By changing the reaction pathway, we do not change the energy relationships between the starting materials and the products. The only energy change is in the height of the barrier – the amount of inertia that prevents the whole system rolling downhill to equilibrium. Consequently, the equilibrium is not changed by altering the reaction pathway.

Of great importance to living as well as to industrial processes are substances called catalysts. Traditionally these are defined as substances which alter the rate of a chemical reaction, but which remain unchanged themselves at the end of the reaction. From this definition, it seems likely that catalysts work by changing the reaction pathway, smoothing out the bump of activation energy. In addition, some also help in ensuring that when a collision between reactants occurs, the collision geometry is right for reaction.

While the good chemist is always conscious of the limitations placed upon him by energy considerations he is usually more interested in making particular products, regardless of cost in terms of money or energy (this is not true in industry, which generally does not bother with reactions which have a substantially positive free energy change), and in finding out what compounds he has made and how they form. Except with simple reactions, he is likely to produce a mixture of products which needs to be separated and identified.

At one time, such separation and identification was a laborious task. Several months were needed to work out the structure of even a simple new compound. That this situation has changed totally accounts for the speed with which many chemical discoveries have been made in the past few decades. Some of the instrumentation that has effected this change is described later in this chapter. First, however, it is necessary to look more closely as what is meant by the 'structure' of a molecule.

## Chemistry in space

Just over a century ago, two European chemists, J. H. van't Hoff and J. A. LeBel, independently developed a theoretical explanation to account for Louis Pasteur's observation that crystals of tartaric acid occur in two distinct geometric forms that are mirror images of one another (like a left and a right hand). In 1875, van't Hoff published an extended version of his theory in French, called *La chimie dans l'espace*. The English translation of this – 'Chemistry in space' – has a contemporary ring to it. However, van't Hoff was not writing about extraterrestrial chemistry, but about the way in which molecules occupy space: that is, their shapes in three dimensions.

Any molecule occupies space in a particular way; that is, it has a three-dimensional structure. On paper, we may write the structure of the water molecule $H_2O$ as H—O—H. But this only indicates how many bonds there are in the molecule, and which atoms they join together. In a water molecule, the oxygen atom has four electrons in its outer shell which are not involved in bonds. However, they affect the disposition of the bonding electrons, because of the repulsive effect of their negative electrical charges. A slightly more accurate two-dimensional representation of the water molecule is:

```
   O
  / \
 H   H
```

with a 105° angle between the two hydrogen atoms.

In the case of a carbon atom which is linked to four other atoms by single bonds, as, for example, in methane ($CH_4$), the repulsion between orbitals is minimized if the structure is tetrahedral; that is, the four substituents occupy the corners of an imaginary tetrahedron, with the carbon atom at its centre. It was by assuming such a model for the carbon atom that van't Hoff and LeBel laid the first solid foundations for stereochemistry, as the three-dimensional structure of molecules is called.

They showed that, where a carbon atom is bonded to four substituents, all of which are different from one another, two tetrahedra can be obtained which are mirror images of one another (Figure 6). An important property of mirror-image shapes is that they are not interconvertible (try putting a left glove on a right hand). Consequently, mirror-image molecules are different chemical substances – like the different forms of tartaric acid. However, most physical methods are unable to distinguish between a left-handed and a right-handed specimen of the same molecule. There are

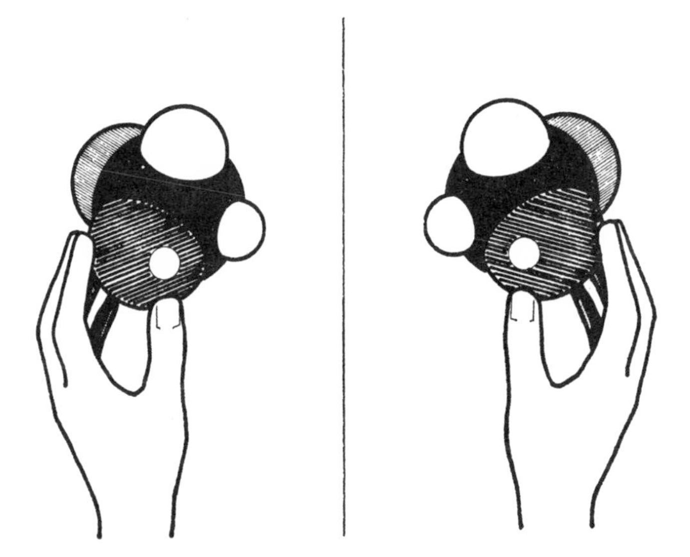

FIGURE 6 When a carbon atom (as an example of a single compound) has four substituents attached, each different from the others, more than one isomeric form is possible. The illustration shows two isomers which are mirror-images of each other (each shaded part-sphere represents one of the atoms and its accompanying electron cloud) and are, therefore, no more superimposable than the hands 'holding' them. Such mirror-images differ in the effect they have on certain forms of light, and are consequently called optical isomers. Although two optical isomers behave in many of the same ways, most biological systems depend on one particular isomeric form.

methods based on their optical properties which can be used to differentiate them. For this reason, molecules which differ only in their handedness are called optical isomers.

Isomers, in general, are molecules which contain the same numbers and kinds of atoms, but in different configurations. In addition to optical isomers, there are structural isomers and geometrical isomers. Structural isomers are compounds in which the atoms are bonded together in different ways. For example, both ethanol (potable alcohol) and dimethyl ether (methoxymethane) are made up from two atoms of carbon, six of hydrogen and one of oxygen. But, in ethanol, the atoms are linked together like this:

```
H     H
 \    |
H—C—C—O—H
 /    |
H     H
```

while, in dimethyl ether, they are linked like this:

```
H          H
 \        /
H—C—O—C—H
 /        \
H          H
```

The differences in the ordering of the atoms produces molecules with widely different properties. Structural isomerism accounts mainly for the very wide variety of possible molecules. For example, it can be shown that there are about 100,000 different possible ways in which nineteen carbon atoms and forty hydrogen atoms can be all linked together.

Geometrical isomerism is slightly more complex and occurs where atoms are held together in closed loops or by multiple bonds. In a compound joined by ordinary sigma bonds, the symmetry of the electron probability between adjacent nuclei makes it possible for them to rotate with respect to each other. In ethane, for example, two carbon atoms are joined together, and three hydrogen atoms are attached to each carbon atom. Each carbon atom is tetrahedral, as we would expect, giving a molecule shaped like that in Figure 7a. If the carbon atoms rotate about the sigma bond joining them, the two sets of three hydrogen atoms at each end of the molecule will act like little windmills. This process requires energy, because the electron clouds around the hydrogens at one end of the molecule will repel the electron clouds around the hydrogens at the other end of the molecule. Repulsion is minimized if the molecule adopts the 'staggered' conformation, in which the hydrogens at one end are out of alignment with those at the other end. This allows the greatest distance between electron clouds. For rotation to be possible, the molecule must be able to pass through the 'eclipsed' form, in which event it has to overcome the repulsion energy of the two electron clouds. Since they are well separated, this is only small – about the same amount as is needed to stretch a hydrogen molecule by 0·1 ångstrōm. However, it has been found with tetrabromoethane ($Br_2HC—CHBr_2$), where two hydrogens on each carbon have been replaced by the much larger bromine atoms, that two

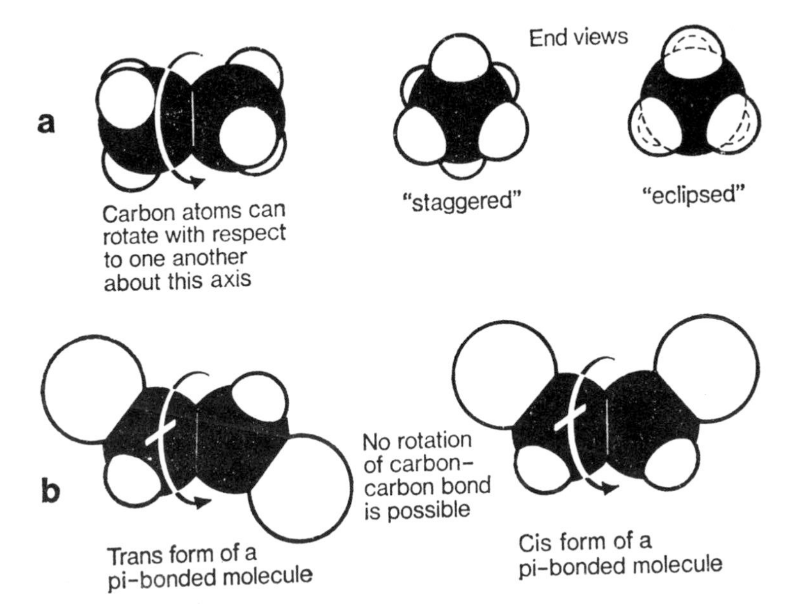


FIGURE 7 (a) Rotation around a single bond in ethane ($C_2H_6$). Both atoms have other atoms arranged tetrahedrally around them but, because the two carbon atoms are joined by a single bond, they can rotate with respect to one another. This means that, when the molecule is looked at from one end, the hydrogen atoms attached to one carbon can adopt different positions with respect to those on the other carbon—the two extremes are the 'eclipsed' form, in which the hydrogen atoms are aligned, and the 'staggered' form, in which they are as far out of line as possible. Because of electron-electron repulsion between the hydrogen atoms attached to adjacent carbons, the 'staggered' form is usually more stable than the 'eclipsed' form. Where the substituent atoms are bulky, this electron-electron repulsion may hinder rotation to such an extent that different forms of the molecule can be isolated. (b) Where two carbon atoms are joined by a pi bond as well as a sigma bond, rotation is prevented. This means that, if the carbon atoms have different substituents, two distinct geometrical isomers (called cis and trans) are isolable.

distinct forms of the molecule can be isolated at low temperatures. This is because the bulkiness of the bromines makes it difficult for them to rotate past one another. More energy is required than is available at low temperatures.

When pi bonds are present the situation becomes slightly more complex, for these bonds are directional in space. The two carbons in ethylene, which are joined by a sigma and a pi bond, can only rotate if the pi bond breaks. Although pi bonds are less energetically stable than sigma bonds, it nevertheless requires a substantial amount of energy to break a pi bond. Consequently, even at room temperature, a compound such as 1, 2-dibromoethene (BrHC:CHBr) occurs in two forms, known as geometric isomers, which can be physically separated and which do not interconvert if kept in separate bottles on the laboratory shelf. Double bonds, or triple bonds (as found in the cyanide ion —C≡N), are in a sense stiffeners of a molecule's backbone.

The 'stiffening' effect of pi bonds may, in certain circumstances, affect more than the two carbon atoms that are joined by the bond. Where more than one pi bond occurs in a molecule, with a single sigma bond separating the two pi bonds, it is possible for electron delocalization to occur, so that the pi electron cloud spreads over all four carbon atoms involved.

Because of the different spatial possibilities when a number of atoms are linked together, knowing the numbers and types of atoms in a molecule gives no clue to its structure. This is why the rise of instrumental chemistry has been so important in the development of the subject during the past few decades. A century ago, it was not possible to work out the structure of any but the most simple molecules. Fifty years ago, the structures of more complex molecules could be worked out in greater detail, but only by lengthy procedures which often consumed substantial quantities of material in the process. A chemist who has worked hard to make a small amount of some new substance does not want to see it used up in analytical tests to determine its structure. Fortunately, the rise of instrumental chemistry now allows most chemists to determine structures, sometimes in minutes, often in days, although sometimes still in months, with small quantities of materials. What has happened to make today's chemical laboratories so different in their methods and techniques from laboratories before World War II?

## New instruments for chemical tunes

The major breakthroughs have been in spectroscopic techniques, which may be divided into several distinct types. All these techniques rely on the interaction between matter and some portion of the electromagnetic spectrum. The electromagnetic spectrum is merely a blanket phrase to cover all those waves that travel at the speed of light and do not need a medium for their propagation; that is, they can travel through a vacuum. Table 2 shows the main subdivisions of the electromagnetic spectrum, from X-rays at the high-energy end to radio waves at the low-energy end. All these forms of electromagnetic radiation are used in chemical analysis today and, because molecules interact differently with radiation of

TABLE 2 *The electromagnetic spectrum*

| | *Wavelength* | *Frequency* (Hz: cycles/ second) | *Energy quantum* (kcal/ mole[1]) |
|---|---|---|---|
| x-rays, neutrons, electrons | 1Å ($10^{-10}$m) | $3 \times 10^{18}$ | 286,000 |
| ultra-violet visible | 1000Å ($10^{-7}$m)<br>1$\mu$m ($10^{-6}$m) | <br>$3 \times 10^{14}$ | 286 |
| infra-red | 10$\mu$m ($10^{-5}$m)<br>100$\mu$m ($10^{-4}$m)<br>1mm ($10^{-3}$m) | <br>$3 \times 10^{12}$ | 2.86 |
| microwave | 1cm ($10^{-2}$m)<br>1m | $3 \times 10^{10}$<br>$3 \times 10^{8}$ | $2.86 \times 10^{-3}$ |
| radiowave | 10m<br>100m | <br>$3 \times 10^{6}$ | $2.86 \times 10^{-6}$ |

1. A 'mole' is a traditional chemical quantity, containing approximately $6 \times 10^{23}$ particles.

different energies, a great variety of information can be obtained from them.

As mentioned in the penultimate section electrons in a molecule can absorb energy and move into an excited state (p. 43). The precise amount of energy which any electron may absorb depends on its environment – the type of molecular orbital it occupies, for instance. Consequently, by seeing what particular wavelength(s) of radiation are absorbed by a molecule, it is possible to work out aspects of its structure.

Ultra-violet radiation is sufficiently energetic to excite many electrons, and an ultra-violet spectrum can be measured in a number of ways. The most general is to pass ultra-violet radiation through a sample of the chemical being studied, and determine how much of the radiation is absorbed. By using scanning spectrometers it is possible to vary the wavelength over a range, and thus detect which wavelengths are most strongly absorbed. This would be the method of choice for examining a compound of unknown structure. By comparing the wavelengths of maximum absorption with reference data, it is possible to suggest particular groupings of atoms which seem likely to be present.

Although the spectroscopies, with some exceptions, are not new, their rapid development since the end of World War II has been dependent on two factors: advances in instrument manufacture, and increasing availability of reference data correlating particular aspects of molecular structure with particular absorptions. Determining the ultra-violet spectrum of an unknown compound would be useless if the absorptions at different wavelengths could not be identified by reference to data previously collected by examining the spectra of a wide range of known compounds.

Pushing an electron from one energy level up to another is not the only way in which radiation may affect a molecule. Less energetic radiation, such as that from the infra-red region of the spectrum, may also be absorbed by molecules. In this case, the energy is absorbed by molecular movements, such as the stretching and rotation of bonds (see p. 42).

In 1953, Professor Bryce Crawford, Jr., of the University of Minnesota, wrote: 'Infra-red spectroscopy has grown like a mushroom in the past ten years. Before the war it was employed by only a few chemists and physicists, using home-built or custom-built infra-red spectrometers. Now the instrument is a standard commercial item supplied by a competitive industry to chemical and medical researchers all over the country (U.S.). More than 1,300 commercial infra-red spectrometers, each representing an investment of two to six Cadillacs, are earning their way in scientific laboratories and industrial plants.'[2] The mushroom has continued growing; you would now have to trade in a handful of infra-red spectrometers before getting a Cadillac in exchange. But there are far more to trade in. Most chemical laboratories have several, generally ranging from a simple routine model, which an undergraduate student can learn to use in half an hour, to complex machines designed for advanced research into infra-red spectroscopy itself. The latter are gradually pushing back the wavelength frontiers to the limits of the infra-red region of the spectrum, making the technique useful for a wider range of chemical problems.

Since the development of infra-red spectroscopy in the 1930s, an enormous number of spectra of known compounds has been catalogued and, from this, it has been possible to isolate particular groupings with characteristic absorptions. For example, the ketone or aldehyde group, $C=O$ (carbonyl), usually absorbs strongly radiation of between about 0·57 and $0{\cdot}59 \times 10^{-5}$ metres wavelength. No exact absorption wavelength can be given, because the remainder of the molecule interacts with its carbonyl function, and such interactions shift the carbonyl absorption up or down the scale by a small amount. However, by comparing a spectrum of an unknown organic compound with a table of absorptions, it is possible to gather numerous clues about its molecular structure.

2. Bryce Crawford, Jr., 'Chemical analysis by infra-red', *Scientific American*, October 1953, p. 42

A chemist studying an organic compound of unknown structure will find a scanning infra-red spectrometer most useful, quickly showing him the major absorption frequencies over a wide range. For following a particular reaction, either infra-red or ultra-violet spectroscopy at a single wavelength can be used. If, for example, we are studying the kinetics of a reaction (that is, the rate at which it occurs and the factors on which it depends), and we know that one of the products has a strong infra-red or ultra-violet absorption at a particular wavelength which is not present in any of the starting materials, it may be possible to perform the reaction in a vessel linked to a spectrometer set at that specific wavelength, and thus follow the build-up of product by the increase in intensity of absorption. Intensity of absorption can be related to concentration by using simple equations, thus allowing the actual amount of product formed to be calculated.

If a chemist is performing an unknown reaction, he can follow the disappearance of starting materials by the disappearance of a characteristic frequency. Such techniques are useful for checking qualitative progress. For example, suppose that a ketone is being converted to some other compound which contains no carbonyl group. If the reaction has not been done before, the chemist will have little idea as to how long it will take or where the equilibrium lies. But if once every half-hour he removes a tiny portion of his reaction mixture – only the smallest drop is required – from the reaction flask and quickly runs an infra-red spectrum on it, he will be able to see whether the carbonyl peak is disappearing satisfactorily. If a noticeable decrease takes place at first and then stops, he can guess that equilibrium has been reached and there is no point in continuing the reaction. He can now set about separating his products from the residual starting materials. Among these separation processes, he will probably use at least one that was not available thirty years ago.

## Molecular radio

The interaction of radio frequencies with molecules was noticed in the 1930s, but it was the extensive development of radar during World War II that provided the expertise required to make spectroscopy in this part of the spectrum feasible. The earliest to be developed was microwave spectroscopy.

Microwave spectroscopy, which utilizes wavelengths between 1 mm and 10 cm, measures the rotational spectra of molecules, and is restricted to fairly simple molecules. As mentioned earlier, we can imagine a diatomic molecule as a dumb-bell, the ends of which rotate about the centre. Altering the speed of rotation requires energy in specific amounts, which can often be provided from the long infra-red or microwave regions. Microwave spectroscopy provides information about bond lengths and, in more complex molecules such as sulphur dioxide, bond angles. Although this technique is not applicable to problems involving the extremely complex molecules of interest to many chemists, it is of use to the physical chemist in making precise measurements on simple molecules and in providing some information not available from other sources. In common with several other analytical techniques familiar to the chemist, such as Raman spectroscopy, its range of utility is expanding rapidly as a result of the increasing availability of lasers.

The compounds formed between small organic chemical groups and certain metals are not open to complete structural analysis by infra-red or ultra-violet spectroscopy; however, microwave spectroscopy can sometimes provide the clues necessary to deduce their structures. Microwave spectra can be used also to pinpoint unusual isotopes in a molecule. The most common form of carbon is carbon 12, which has six neutrons and six protons in its nucleus. Another isotope, carbon 13, has an extra neutron. Obviously, this makes the atom slightly heavier, and affects the energy needed to rotate a molecule containing it. In addition, the energies will differ slightly according to the location of the heavy carbon atom in the molecule. Thus, microwave spectroscopy can be used to identify molecules of methylacetylene with a $^{13}C$ atom in the methyl group ($^{13}CH_3C{\equiv}CH$) as different from those in which the $^{13}C$ is the middle carbon atom in the molecule ($CH_3{}^{13}C{\equiv}CH$).

More important to the chemist trying to work out structures of complex compounds are forms of spectroscopy which combine the behaviour of molecules in magnetic fields with their interaction with radiowaves or microwaves. Some of these techniques were made possible by the immense technical development of radio and microwave generation methods during work on radar systems in World War II, but the actual effects involved were not discovered until after the war. In 1946, building on the basis of prewar work on nuclear magnetism by the Nobel prizewinning physicists Otto Stern and I. I. Rabi, Edward Purcell and Felix Bloch discovered the nuclear magnetic resonance (nmr) effect. They were awarded the Nobel physics prize jointly for their discovery in 1952. Like a number of other physicists' discoveries, this proved of more benefit to chemists as a practical tool.

Basically, any element which has an unpaired proton in its nucleus will behave like a small bar magnet. In a magnetic field, the nucleus can take up either of two orientations: one aligned with the magnetic field, the other opposed to the magnetic field. The second state has the higher energy and, by passing microwave or radiowave frequencies through molecules containing such nuclei, it is possible to flip the aligned nuclei into the higher energy state. Not many common nuclei have unpaired protons (a nucleus with an odd number of protons is not the same; from a nuclear point of view, a proton can pair with a neutron). Fortunately for the organic chemists, hydrogen, one of the elements most common in the compounds with which they deal, does have an unpaired nuclear spin.

Proton magnetic resonance – nuclear magnetic resonance concerned solely with hydrogen – can provide the chemist with much information about the environment of hydrogen atoms in a molecule, for the surroundings – how electron clouds are spread around in nearby orbitals – slightly affect the interaction between the applied magnetic forces and any particular hydrogen atom. As a result, differently placed hydrogen atoms in a molecule have slightly different flipping energies. From the 'perturbations' – energy changes caused by 'shielding' or 'deshielding' of a particular nucleus – it is possible, for example, to tell the difference between a carbon atom with three hydrogen atoms attached to it and one with only two. As the spectrum also gives an idea of the concentration of particular species, it is possible to solve points of fine structure that are ambiguous in infra-red studies.

A few other nuclei have the required unbalanced spin for nuclear magnetic resonance; fluorine is one example. It is also possible to use unusual isotopes of nuclei that commonly do not have unpaired spin. The heavy carbon isotope $^{13}C$ is widely used for advanced nuclear magnetic resonance studies.

In addition to unpaired nuclei, unpaired electrons can also flip in a magnetic field to give electron spin resonance (esr) spectra. Unfortunately, most chemical compounds do not have unpaired electrons. However, if a molecule with paired electrons breaks up, so that a two-electron bond is broken evenly, one electron goes to each part of the molecule, and both fragments, called free radicals, will then have unpaired electrons. This is not the only way bonds break. Both electrons may go to one fragment, making it negatively charged and leaving the other fragment positively charged. However, unpaired electrons are found sufficiently often to make esr a useful addition to the chemist's armoury of instruments.

The use of magnetic characteristics of molecules is not restricted to nuclear magnetic and electron spin resonance spectroscopies. If a molecule is bombarded with an energetic electron, it is possible to knock off one of the molecule's own electrons, making the molecule positively charged.

If such a positively charged molecule is placed in a magnetic field it can be made to travel in an arc, the exact shape of which depends on the mass:charge ratio. By using electrons with about seven times the energy required to knock an electron from a molecule, it is possible to produce a number of charged fragments – some of the excess energy goes into breaking chemical bonds – with different masses. These can be detected in a number of ways and, for a particular molecule, will give a characteristic pattern of fragments.

Known as mass spectrometry, this technique is widely used for determining the structures of complex organic molecules. As the molecule has to be vaporized, it is not widely applicable to inorganic compounds or to very large molecules which are not easily vaporized. This is a temporal judgement, for the usefulness of the technique is leading to extensive studies of means for preparing volatile derivatives of involatile materials. A new technique for volatilization, utilizing an isotope of the transuranium element californium, was reported early in 1975. Developed by scientists at Texas A & M University, it uses fission fragments from Cf-252 to create localized, but very high temperatures in a non-volatile analytical sample. This results in the production of quasimolecular ions which are then accelerated down a tube to an ion detector.

Research aimed at broadening the scope of mass spectrometry to include non-volatile materials has been especially intense since methods have been developed for analysing complex mass spectra by computer. Peptides, small fragments from proteins but still highly complex molecules in themselves, are now successfully being analysed by this technique. Another technique developed during the 1960s, when computer analysis of mass spectra began, is the combined use of mass spectra and gas chromatography.

Gas chromatography is one of the separational methods that has revolutionized the chemical study of complex mixtures in the past two decades. Coupling it with mass spectrometry has produced a very powerful tool, useful for such purposes as analysing minute samples of rock brought back from the Moon for traces of the chemicals of life.

Gas chromatography was invented by two British scientists, Archer Martin and Tony James, in 1952, the same year that Martin shared the Nobel chemistry prize with R. L. M. Synge for their discovery during World War II of paper chromatography, a similar but less versatile separational technique. Both of these techniques go back to a nineteenth-century Russian chemist, Mikhail Tswett, who found that a solution of natural pigments (from grass, for example) could be separated into different, coloured bands by adsorbing it on the top of a column of powdered solid, and gradually pouring through it a liquid in which the pigments were partially soluble.[3]

All chromatography works on the basis underlying this phenomenon. Different molecules have different affinities for various solids and liquids. An obvious example is the lack of affinity between oil and water: they do not mix with one another. A mixture of similar molecules may dissolve in the same liquid, and be adsorbed on the same solid. But the strength with which the different compounds are held by solid and liquid will differ slightly. This means that, once adsorbed, they can be desorbed at different rates, and will separate.

A simple example is to take a strip of white blotting paper and place a spot of black ink (fountain or felt-tip pen ink, not ballpoint ink) about half-way along the strip. Dip one end of the strip in a glass of water and let the rest of the strip hang downward. Water gradually travels along the strip. As it passes the ink mark, the spot begins to streak. By the time the water gets to the end of the paper, you may be able to see that the 'black' ink is made from a mixture of different colours. It was a highly refined form of such paper chromatography that won Martin and Synge their Nobel prize.

In gas chromatography, a column packed with a solid coated with a special, high boiling-point liquid, is used in place of the paper; instead of water, the adsorbed compounds are removed by a stream of gas. When they reach the end of the column, which may be at quite a high temperature, they are detected in a variety of ways, and the composition of the mixture can be ascertained. For example, wines and beers have been analysed by gas chromatography to uncover the secrets of their individual flavours. As chemists know most of the compounds likely to appear in such a mixture, individual compounds can be identified by comparing the retention times (how long a compound stays on a particular type of column at a particular temperature) published in the scientific literature with the retention times recorded in the experiment.

If we want to find out the ingredients of a complex mixture of hitherto unknown compounds – isolated perhaps from a newly discovered plant in the South American jungles – chromatography will separate the mixture, but cannot identify the individual compounds as they emerge. Mass spectrometry, on the other hand, can be used to indicate likely structures for pure compounds, but not for mixtures. The answer is called combined

3. Chemists make a distinction between the similar words absorption and adsorption. When a substance is absorbed in a solid, it permeates the whole solid. When it is adsorbed, the substance occurs only on the surface of the solid.

gc-ms: a gas chromatograph with a mass spectrometer fitted on to it. As both techniques need only tiny quantities of material, this marriage of instruments is ideal for analysing rare samples that cannot easily be replaced, such as samples of new plants or of Moon-dust. It is also becoming a powerful tool in clinical biochemistry where it is used for the study of metabolic diseases through analysis of trace compounds in urine samples.

To be analysable by gas chromatography, a compound must be volatile, as is the case with mass spectrometry. In that sense, both types of analysis are compatible. But neither is very useful for the analysis of large molecules, such as proteins. There are techniques for isolating and identifying the structures of such large biological macromolecules, some of which are described later (Chapter 4), when recent work on biological chemistry is discussed in more detail.

### Make way for a gamma-ray

A very recent addition to the spectroscopic work force is another example of experimental physics being appropriated by the chemists. During the late 1950s, Rudolf Mössbauer was studying aspects of the behaviour of $\gamma$-rays at the Munich Institute of Technology in Germany. A few months after he finished his doctoral work, during the summer of 1958, he first observed the Mössbauer effect. If a nucleus emits $\gamma$-rays, these usually have a range of energies, because the nucleus itself recoils as the $\gamma$-ray is emitted, just as a rifle recoils when a bullet is fired. However, if the nucleus is embedded in a solid matrix to prevent recoil, all the $\gamma$-rays are emitted with a closely defined energy. If such a $\gamma$-ray collides with another nucleus of the same element, it is absorbed. This resonant absorption is highly specific; consequently, it was not until recoilless emission was developed that the Mössbauer effect was noted. This effect has been useful in physics – it helped confirm some of Einstein's views on relativity – and Mössbauer was awarded the 1961 Nobel prize in physics for the discovery. Now it is proving useful in chemistry.

Like the shielding effects of the molecular environment in nmr, the structure surrounding a Mössbauer absorber alters the energy absorption level slightly, and thus the energy of the $\gamma$-rays which it can absorb. The $\gamma$-rays can be produced over a range of energies by using another physical principle, the Doppler effect. In practical terms this means that the $\gamma$-ray target is moved away from the $\gamma$-ray emitter at different speeds, until resonant absorption is achieved. From the speed of movement it is possible to calculate the extent of the Doppler effect, and the change in the $\gamma$-ray's energy between emission and absorption.

About half of the studies of Mössbauer absorption have involved compounds of iron. For example, using Mössbauer spectra, it was possible in 1966 to work out the kinetics and mechanism of the breakdown of iron oxalate compounds, a problem which had puzzled chemists since the turn of the century. More recently, the technique has been used for analysing iron in Moon rocks brought back to Earth by the *Apollo* astronauts, for estimating the temperatures at which medieval pottery samples were fired, and for distinguishing apparently similar glazes on Chinese ceramics.

## The X-ray eye

The most accurate technique for structuring an unknown molecule, whatever its size, is not spectroscopic, but a diffraction method. However, like the spectroscopic methods, it relies on the interaction between matter and part of the electromagnetic spectrum. X-ray diffraction was discovered in the early years of this century by two British scientists, W. H. Bragg and his son, W. L. Bragg. Rather than being absorbed by matter, X-rays are diffracted, that is, bounced off. If a beam of X-rays is aimed at a crystal, it is possible to work out the crystal's structure from the pattern of diffracted X-rays as shown on a photographic plate.

Like mass spectrometry, X-ray diffraction did not come into its own as an analytical technique until the development of computer programmes for interpretation of the data obtained. It has subsequently been possible to work out the precise three-dimensional structures of complex molecules such as enzymes and, from their shapes, to postulate various ideas about how they act as catalysts. Neutrons and electrons can be used in place of X-rays in diffraction experiments. The powerful neutron sources required for neutron diffraction are available only in nuclear reactors, so this technique is still limited in its applications; however, the method provides a useful adjunct to the more conventional X-ray technique. Electron diffraction is usually applied to gases, and is consequently limited to use with gaseous materials, where it comes into competition with microwave and rotational infra-red spectroscopy.

From the viewpoint of the chemist attempting to determine a structure accurately, there is no doubt that X-ray diffraction gives the most complete answer. It draws a three-dimensional picture of the molecule that shows not only bond lengths and angles, but also specific stereochemistry. Unfortunately, it still takes several months for most complete X-ray analyses and, for most purposes, the chemist will satisfy himself with one or a combination of the spectroscopic tools at his disposal. Additionally, he will use one of the now generally available techniques of optical analysis, such as circular dichroism or optical rotatory dispersion, if he is studying molecules that have asymmetric centres (those that can exist in optically isomeric forms).

Other analytical methods of great sophistication are also in use. In recent years, methods have been developed for studying the surfaces of materials in atomic detail. Such studies, which are important, for example, in understanding how metallic catalysts operate, have their own specialist techniques. For instance, X-rays can be used to dislodge electrons from the innermost orbitals of atoms. However, they will do this only from atoms in the top 100 ångströms ($10^{-8}$ metre) of a solid. By studying the spectroscopic behaviour of the ejected electrons, or of the atoms, as other electrons drop into the vacated orbitals, it is possible to obtain useful information about surfaces or surface films – which may differ considerably, both in structure and properties, from bulk material.

## Watching reactions at work

Not only can a chemist identify the products of a reaction with the varied methods just described, he can also follow the kinetics in many cases, and possibly establish the mechanism of the reaction from spectroscopic evidence. For example, by a simple reaction between acetone, which is an organic compound containing a carbon doubly bonded to oxygen, and also having two other carbon atoms directly attached to it, and hydroxylamine ($NH_2OH$), an oxime is formed, with the place of the doubly bonded oxygen being taken by a doubly bonded nitrogen.

$$CH_3\overset{\overset{\displaystyle O}{\|}}{C}CH_3 + \begin{matrix} H \diagdown \\ \\ H \diagup \end{matrix} NOH \rightarrow CH_3\overset{\overset{\displaystyle NOH}{\|}}{C}CH_3 + H_2O$$

On paper this looks like a simple removal of the oxygen and replacement by nitrogen. But we must not forget that, in reality, this replacement involves breaking two existing bonds and forming two new ones. If we examine the reaction spectroscopically, we can detect both starting material and product by characteristic absorption of infra-red radiation. In some cases, it is possible for the characteristic absorption of starting material to disappear completely before any sign of the absorption characteristic of the product appears. At this stage, it would seem, the mixture contains neither starting material nor product. This is the case; an intermediate compound – not a reaction intermediate, for two separate reactions are taking place – is formed. The first of the two reactions involves the breaking of one of the two bonds to oxygen and the formation of a single bond to a hydroxylamine molecule's nitrogen atom. As part of the process, the hydroxylamine molecule loses a hydrogen atom, which forms a bond with the oxygen atom, which now has a spare electron. The result looks like this:

```
    OH
    |
 CH3CCH3
    |
 H—N—OH
```

The reaction by which this is formed occurs rapidly; a slow reaction now takes place in which this molecule loses the elements of water to form the oxime product.

Spectroscopic methods are not the only ones used to study reaction mechanisms. Experience in the handling of unfamiliar isotopes of common elements has been of help in elucidating reaction mechanisms. For example, a common class of organic compounds is the esters. Esters can be hydrolyzed to form an acid and an alcohol in two possible ways: by the breaking of either of the two carbon-oxygen bonds (see Figure 8). During hydrolysis, the elements of water add on, so one could imagine H from $H_2O$ joining the oxygen atom whose bond has just broken, and OH joining the carbon atom whose bond has just broken. But how can we determine which bond has broken?

By using water containing an uncommon isotope of oxygen, $^{18}O$ (common oxygen is $^{16}O$), it was possible to find out which of the two products – acid or alcohol – the OH portion of the water molecules latched on to, and consequently to identify the point of breakage of the carbon-oxygen bond. As oxygen 18 is not radioactive, to find out which products contained the heavy oxygen it was necessary to fall back on spectroscopy – in this case mass spectroscopy – to complete the experiment.

This is a simple example of the use of isotopes in elucidating reaction mechanisms. Isotopes can also be used for working out the rates of very slow reactions. The amino acid alanine, found in most proteins, breaks down very slowly in aqueous solutions, releasing carbon dioxide as it does so. By keeping alanine solutions, in which part of the alanine has been 'labelled' with the radioactive carbon isotope $^{14}C$, for several weeks, then blowing out the apparatus with ordinary carbon dioxide, it is possible to remove any radioactive carbon dioxide formed and measure it accurately.

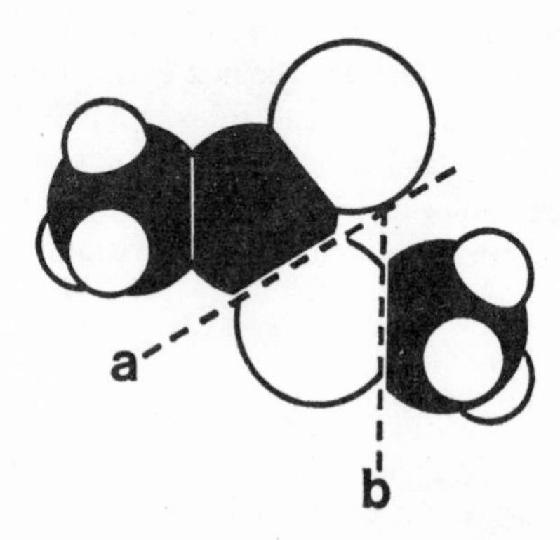


FIGURE 8 An ester molecule can be hydrolyzed by a water molecule in two ways: either carbon-oxygen bond (a) or carbon-oxygen bond (b) might break. The use of 'heavy water' containing the uncommon isotope $^{18}O$ enabled experimenters to find out which of the bonds break under different conditions.

This procedure has shown that, in many aqueous solutions, it would take about 4,000 years for half the alanine present to decompose.

Labelling experiments are based generally on the assumption that a heavy or light isotope of an element will behave in the same way as the most common one. In some cases, because different masses are involved, this is not true, and the difference can also be exploited. This is particularly the case with hydrogen, for its isotopes, deuterium and tritium, are almost two and three times as heavy as normal hydrogen, which means that the strength of a bond between hydrogen and another element is different from that between the other element and either deuterium or tritium. If a reaction involves the breaking of a bond to hydrogen, a proof may be possible by checking the effect of substituting deuterium or tritium into the molecule in place of the particular hydrogen atom under investigation.

### Quick-mix chemistry

Although tracer methods can be useful in determining the rates of very slow reactions, in the last twenty to thirty years chemists have become interested in unravelling the mechanisms of very fast reactions. The advances in this field are truly amazing: it is now possible to study what happens during a few nanoseconds in a fast reaction, and a nanosecond is one thousand-millionth of a second. As chemists don't move fast enough to remove samples from such reactions from their reaction vessels, place them in a spectrometer, and analyse them, how is the study done?

Until special methods were developed, it was not possible to study reactions which occurred in less than a few seconds. One of the problems was bringing the reactants into contact with one another. It was the solution to this during the 1920s that led to the first breakthrough. By keeping reactants in separate solutions, and moving these steadily down the twin arms of a Y-shaped tube, it is possible to get the reaction to start where the arms of the Y meet. By adjusting the speed of the solution down the tube and placing a measuring instrument a specific distance down the leg of the Y, the state of the mixture at, for example, one-hundredth of a second after mixing can be recorded.

If the speed of travel of the solutions is kept constant, and new solution added all the time, measurements made at the downstream point need not be done quickly, for the mixture passing the observation window always represents the reaction after a certain time. Consequently, the reaction stage can be studied at leisure; by speeding up or slowing down the flow, the point will see the reaction at an earlier or later stage. In this way, reactions which are over in a few parts of a second can be studied. By 1940 the American chemist Britton Chance had reduced the actual mixing time for this technique to one ten-thousandth of a second. The technique has been used, for example, for measuring the speed at which haemoglobin, the important respiratory pigment of blood, gives up oxygen which it has picked up on its travels through the lungs (half the oxygenated haemoglobin in a sample deoxygenates in 0·015 seconds.)[4]

When we study the mechanism and kinetics of reactions, it should be remembered that, although we talk of converting reactants to products, it is more correct to talk about reaching an equilibrium of reactants and products. Another way to study fast reactions is to allow them to reach equilibrium under one set of conditions and then change these suddenly, for example by discharging an electric current through a solution, which may raise the temperature by 10 °C in about one hundred-thousandth of a

4. For many processes, a simple rate law operates: half the amount of reactant present changes into product in a fixed period of time. The importance of this law is that it makes no difference when we start measuring, for whenever we start, and however much material we start with, we know that half of it will be gone after a fixed period of time which is called, not surprisingly, the half-life. Spontaneously radioactive elements decay in this way. If a piece of rock contains 10 g of a radioactive element with a half-life of ten years, at the end of ten years only 5 g of the radioactive material will remain. After twenty years (another half-life period), half the remaining amount will have gone, leaving only 2.5 g, and so on. In the haemoglobin case, given 1,000 molecules of oxygenated haemoglobin, 500 of them will shed their oxygen within 15 milliseconds—after one-tenth of a second there would be less than 15 molecules of haemoglobin with oxygen still bound to them.

second. The rate at which the mixture comes to its new equilibrium point can then be measured by, for example, spectroscopic techniques. Relaxation methods, as such sudden changes are called, were introduced in 1954 by the German chemist Manfred Eigen, who shared the 1967 Nobel chemistry prize for developing them.

One of the earliest reactions studied by Eigen and his colleagues was the combination of hydrogen and hydroxyl ions to form water molecules. The result indicated an enormously high rate – higher, in fact, than the calculated rate of collisions for hydrogen and hydroxyl ions under the conditions used. Can a reaction occur without a collision? The answer is no. The calculation was wrong, for the reaction does not involve collisions between hydroxyl ($OH^-$) and hydrogen ($H^+$) ions. These two ions go about in association with water molecules, and it was found that the reaction between hydrogen and hydroxyl takes place under the guise of two more complex aggregates coming together, $H_9O_4^+$ and $H_7O_4^-$ (one ion surrounded by four, the other by three, neutral water molecules). When the collision rate of these was calculated, it gave the answer achieved experimentally by Eigen.

Another fast-reaction technique is flash photolysis. Its discoverers, Reginald Norrish and Sir George Porter, both British chemists, shared the 1967 Nobel chemistry prize with Eigen. Flash photolysis began life in Cambridge in 1949, and has since proved its value again and again. The basic principle is to initiate a photochemical reaction (one in which the necessary energy is supplied by light) with an intense, controlled flash of light which is complete in a very short time, and to follow this, after a specified interval, with a weak beam of electromagnetic radiation which measures any spectroscopic changes brought about by the earlier flash.

A large number of chemical systems have been studied in this way, ranging in complexity from the recombination of iodine atoms into molecules to the photochemical reactions involved in such complex biological processes as photosynthesis. The development of lasers during the 1960s, and their application as a light source for the flash, has extended the technique to an examination of shorter and shorter time scales. As Porter pointed out in his Nobel speech, 'without special techniques, [man] is limited in his perception to times between about one-twentieth of a second (the response time of the eye) and about $2 \times 10^9$ seconds (his lifetime)'.[5] Working in the picosecond ($10^{-12}$ seconds) region, chemists are now observing events that are over in one ten-thousand-millionth of the eye's response time. Currently, there is excitement in the photochemical field

5. George Porter, *Nobel Lectures, Chemistry* **1963–70**, Elsevier, London, 1972, p.241

about a new technique for studying photochemical reactions which combines flash photolysis, electron spin resonance spectroscopy and a sophisticated variant of NMR spectroscopy (called chemically induced dynamic nuclear polarization). Because of the depth of analysis this gives it may soon be possible to design photochemical experiments to produce desired products far more effectively than has been the case in the past. Such a fundamental breakthrough could have a major effect on the industrial processes used for producing some chemical products.

Although many chemical reactions occur in microtime, and need advanced techniques – lasers, oscilloscopes, spectrometers, and so on – for their study, analysis of chemical reactions still relies on the human brain. Instruments only extend our senses; they do not replace our intelligence. Intuition is still important for the chemist interested in elucidating how a reaction occurs, for unless he has an idea about what is happening, he is unable to plan the crucial experiment that will tell him whether he is right or wrong. Nevertheless, it is important to realize that few of the results described in the remainder of this book could have been achieved without at least some of the techniques described in this chapter.

# *CHAPTER 3*

## CHEMISTRY IN THE BEGINNING

The chemical principles described so far are not of theoretical interest alone, with no application to everyday life. They underlie the shaping of the world we live in, from our synthetic fibre clothes right back to the cataclysmic upheavals at the dawn of terrestrial time. Although the early history of the Earth can be seen in chemical terms, is such a vision new? In a genuine sense, the answer is yes. Not only is one of the most valuable techniques for studying the age of objects – radioisotope dating – a development of postwar chemistry, but a modern knowledge of chemistry also includes the behaviour of compounds at high temperatures and pressures – conditions only recently obtainable in the laboratory, but likely to have obtained on the Earth during its development. Thus, a knowledge of modern chemistry adds new evidence to the arguments of geologists about the Earth's formation and history.

To begin at the very beginning, it has been estimated from physical evidence about the distance of galaxies from the solar system, and the speeds at which they are moving away from us, that the universe is approximately fifteen thousand million years old. On the basis of chemical studies of different isotopes of particular elements in meteorites, it has been estimated that these elements were formed about eleven thousand million years ago.

Some chemical elements decay radioactively to produce gaseous elements. If, during this decay, the elements are locked up in a solid, cold body, the gaseous products will be trapped. It is possible, by measuring the amount and type of trapped gas, and the residual amount of the elements from which it has formed, to calculate how long the bodies have been solid. Such radioisotope dating of meteorites indicates that they reached their present form about four and a half thousand million years ago. This is the time when it is believed that the solar system took on a recognizable structure. There are, however, no rocks on earth which give radioisotope dates as old as those obtained from meteorites. This indicates that, for some time after its formation, the Earth was geologically in a state of intense activity for, if a rock melts, it loses trapped gases, and the radio-clock is set back to zero. The oldest unaltered rock samples on earth are only about 3·9 thousand million years old.

How the different bodies in the solar system – Sun, planets, meteorites, and satellites – formed is still disputed. According to Jon Darius, a researcher at the University of London Observatory, 'there are as many hypotheses as there are investigators – indeed more – and . . . far from approaching a gradual consensus, their views are diverging more today than ever before', while a correspondent to the scientific journal *Nature* suggested more recently that one should not refer to a group of cosmogonists because 'a "nebulosity" of cosmogonists might be the better collective noun considering the scant common ground between their theories'.[1]

One view which seems to account for what is known currently about the chemistry of the Earth and the other planets in the solar system is a modern variant of an idea put forward by the Marquis de Laplace in 1796. He suggested that, at some time in the past, a large solar nebula contracted, gradually throwing off rings of material as it did so. Each of the planets formed from one of these rings.

For many years this theory was in disrepute, but it has now been revitalized by modifications which negate the criticisms raised against it. The 'missing link' from Laplace's original hypothesis is, according to Dr A. J. R. Prentice of Monash University in Australia, 'supersonic turbulent convection'. Among other things, this type of motion within the young Sun would have led to a dense ring of gas forming at its equator. Prentice has shown that, using his variant of Laplace's hypothesis, if one calculates where such rings would have been left behind by a contracting Sun, their orbits correspond well with those of the present-day planets.

The chemical consequences of Prentice's work have been studied by Professsor A. E. Ringwood of the Australian National University. Except very close to the Sun, he argues, the protosolar system was a cold place. The rings left behind by the contracting Sun would have been at a temperature less than the freezing point of ice during planetary formation.

Ringwood envisages rings around the Sun in which rock and ice fragments circulated at the centre, surrounded by a more diffuse tube of gaseous materials, predominantly hydrogen and helium. 'Ice' here does not mean just frozen water, but a variety of substances which would have been solid at the ring temperature, but are liquid or gaseous on the present-day Earth. Under these conditions, iron would have been present only in its oxidized forms.

Gradually, solid particles would have come together (accreted) under

1. Jon Darius, 'Cosmogony now', in *New science in the solar system* (ed, Peter Stubbs), I.P.C., London, 1975, p. 2; David W. Hughes, 'Solar system origin', *Nature*, cclx, 1976, p. 15.

gravitational attraction, releasing gravitational potential energy thermally as they did so. This would have occurred slowly at first, and the thermal energy would have been dissipated by degassing of substances such as water and methane. As the planetesimals got bigger, and more energy was released through accretion, iron oxides would have reacted with carbon to form free iron and carbon monoxide gas.

In the case of the Earth, Ringwood proposes that, as yet more gravitational energy was dispersed, the proto-Earth became hotter, so that the dense iron sank to the middle through partly molten rocky material. The atmosphere at this time would have consisted not only of carbon monoxide and hydrogen, but also the light metals such as sodium. Eventually, enough heat would have been produced (through radioactive decay as well as the liberation of gravitational energy) to vaporize some of the silicate rocks.

It is possible, according to this scenario, that the Earth's atmosphere then behaved much as the Sun had done, forming a dense ring around the planet. This ring subsequently accreted to form the Moon. Ringwood's hypothesis neatly explains the differences in gross composition between the Moon and the Earth, and makes predictions about the compositions of the other planets which seem to accord with what is known at the moment.

When more hard information about the other bodies in the solar system becomes available, as a result of space exploration, it will be easier to decide which of the many hypotheses about the solar system's early history best fits the facts. A different chemically-oriented hypothesis has been put forward by John S. Lewis of the Massachusetts Institute of Technology. His basic assumption is that a great mass of dust and gas particles surrounded the early Sun, and that it became less dense and cooler the further it was from the centre. He postulates much higher temperatures for the formation of the inner planets than those suggested by Ringwood.

According to Lewis's model, the Earth was formed at about 300 °C, and he accounts for its composition, and that of other planets, on the basis of different chemical equilibria. For example, the high density of the innermost planet, Mercury, can be accounted for by arguing that it formed at too high a temperature for silicate minerals to condense, and is therefore rich in iron. On the other hand, both Earth and Venus would then have condensed in a cooler region where the less dense silicate rocks were stable, thus accounting for the lower densities of these two planets.

The main difference, according to Lewis, between Earth and Venus is that our planet formed at a slightly lower temperature, so that iron sulphide and tremolite (a hydrous silicate) are present on Earth but not, according to spectroscopic observations made by artificial satellites, on Venus. Both these minerals, says Lewis, were crucial in the later history of the Earth.

Lewis's theory envisages melting of the Earth which, he says, would have taken place at a low temperature because of the presence of sulphur, and may have begun before the accretion process was completed. This would have permitted the early formation of an iron core to the planet – a necessary part of any theory of the Earth's evolution, as it is known from studies of magnetism in rocks that the Earth had a magnetic field comparable in strength to the present-day field at least three and a half thousand million years ago.

Equally important, according to Lewis, is the fact that the Earth formed in a region where tremolite could exist. As temperature and pressure decrease, in a mixture of elements of the believed 'solar nebula' composition, tremolite is the first solid to form which contains water. Hence, there is abundant water on Earth today, freed from the minerals to which it was bound by subsequent geological activity, but only traces on Venus.

Possibly neither of these views truly describes the formation of the Earth. On the other hand, perhaps each is true in some respects. Only time, and further extraterrestrial exploration, will tell. Perhaps, in any case, we ought to be more concerned with the Earth as it is now, than as it was in the beginning.

The Earth at present is made up of several layers. In the centre is the solid, inner core, which is probably composed mainly of an iron-nickel alloy. Surrounding this is another, liquid core, made largely of iron sulphide. Outside is the mantle, made of rock and, surrounding this, is a thin crust, basically of a different type of rock, which forms the surface and supports the hydrosphere (water) and atmosphere.

Once the core had formed, and the rocky slag floating atop it had cooled sufficiently, different compounds would have formed, according to chemical laws. As solids appeared in the melt, they would have obeyed physical laws and floated to the surface or sunk, according to their density. Under the extreme conditions prevailing, other processes, such as injection, would have also played a part. In injection, molten droplets or crystals of material would have been forced under pressure into solidifying rock, and would then have been carried with that rock wherever it went. The world's largest deposit of exploitable nickel and platinum metals in Sudbury, Ontario, is believed to have arisen in this way.

In 1949, geochemist H. Neumann briefly summed up how his discipline should work: 'The primary principle of geochemical classification of the elements must obviously be their tendency to form ionic, covalent or metallic bonds.'[2] If one looks at the elements in the light of this dictum, a number of features of the present-day rocks are explicable.

The metallic elements can be broken down into three categories, roughly corresponding to three types of bonding. The first class, called siderophiles, are elements that tend to associate with metallic iron – the ones that form metallic bonds. Gold is strongly siderophile, but why? If we look at its electronic structure, the outermost ring of electrons contains a single electron, just like sodium and potassium, which are highly reactive metals. Why is gold so unreactive that it is found as a free metal in metallic mixes? Of course it has many more electrons than sodium and potassium, and an anomalously high first ionization energy. The ionization energy, as explained in Chapter 1 is a measure of how strongly an electron resists being removed from the element. Normally, as elements get heavier, the ionization energy falls, because the inner rings of electrons shield the outermost electrons from the nucleus and lower its binding effect. Possibly, in gold, the inner rings are inefficient in shielding the outermost electron, so that the nucleus does bind it strongly. Whatever the answer, gold has a high ionization energy, which is a sign of a siderophile.

The other two classes of metallic elements are chalcophiles and lithophiles, which, respectively, favour covalent and ionic forms of bonding. Chalcophiles predominantly form sulphides, and the lithophiles form silicates. There are borderline cases: manganese, for example, is found in both sulphide and silicate minerals. In general, metals can be classed as one of the two types on the basis of ionization energy, which depends on three important factors: screening efficiency (already mentioned), ionic charge and ionic size.

What effect does ionic size have on the formation of the chalcophile minerals? The most abundant rocks in the earth's crust are basalt and granite, one of the principal components of which is K-felspar, a potassium aluminium silicate. Potassium is a relatively abundant element, more abundant than the other elements that appear directly beneath it in the periodic table: rubidium and caesium. There are no mineral deposits containing only rubidium, but all K-felspar contains traces of this element. On the other hand, the rare element caesium is found in a mineral called pollucite, and not as a contaminant of K-felspar. All three cations are singly charged, so, as crystals of K-felspar were formed from a melt containing other elements, it seems logical that some other singly charged cation might jostle a potassium ion out of place every so often. In fact, an element from another part of the periodic table, tantalum, is a contaminant of K-felspar because its most stable cation is singly charged.

2. H. Neumann, *Mineralogical Magazine*, xxviii, 1949, p.575.

Of course, an intruder can only move in if the outgoing potassium cation leaves sufficient room for it. The ionic radii of the elements we are talking about are (in ångströms) potassium, 1·33; rubidium, 1·44; caesium, 1·67; and tantalum, 1·45. Caesium is much bigger than the others, and will not fit into the K-felspar crystal lattice, which is why it has to form a mineral on its own.

Similar arguments can be applied to other families of elements. K-felspar, in common with all aluminium-containing minerals, contains traces of gadolinium, for the gadolinium cation, like aluminium, is triply charged, and its ionic radius is within 20 per cent of that of aluminium (a 15–20 per cent difference in size seems to be as much as a crystal lattice can tolerate). Felspars also contain silicon. Below silicon in the periodic table is germanium, which forms a cation of the same charge and has a 15 per cent larger ionic radius. Not surprisingly, silicon-containing minerals always contain traces of germanium.

In all our arguments so far we have taken a single property at a time, and kept others (such as ionic charge) constant, in an attempt to build simple models that will help us understand the immensely complex chemical system of the early Earth. But we must always bear in mind that, when we change from one element to another, other properties may change and make our model invalid. For example, in the ionic radius/charge model, sodium felspars should always be contaminated with copper, the ionic radius of which is only 5 per cent smaller. This does not happen, because there is more covalent character in a copper-oxygen bond than in a sodium-oxygen bond. Recent attempts have been made in the lab to synthesize minerals that do not occur naturally, but they do not always work. For example, it has not been possible to synthesize the copper analogue of the common sodium aluminium silicate, albite (Na-felspar).

We also must not forget that, although at some time on Earth the formation of a particular mineral was favoured, conditions may have changed – perhaps on a local rather than a global scale – and altered the mineral structures. Sodium analcite, for example, is a mineral composed of oxides of sodium, aluminium, silicon and hydrogen (i.e., water). If heated very strongly, sodium analcite turns into albite. On the other hand, if synthetic Rb-felspar is heated, it turns into Rb-analcite. This is because analcite has a more open crystal structure than the felspars; in Na-analcite, the water and sodium ions are used to fill space in the crystal structure. Because of their size, they are not very effective fillers, and under the influence of heat the structure collapses to the more compact felspar structure. Conversely, rubidium has very little shoulder room in the felspar structure. Given the heat necessary for the change, it opts for the more open analcite structure. The larger caesium cation, as we have already mentioned, will not fit into the felspar structure at all, but its analcite structure, the mineral pollucite, is very stable.

Gradually, through a closer look at the structure of cations and their properties, and the results of mineral synthesis, we are building up a picture of how the Earth may have developed. The Earth, as we see it, has of course been extensively reworked, both physically and biologically. Many of today's minerals are the results of biological action – not only the white cliffs of Dover, but also many deposits of sulphides and iron oxides. In addition, through the action of plate tectonics, new material from the mantle wells up through the crust in various parts of the Earth, while, in others, crust disappears down into the mantle again. In some cases, it is possible to identify zones where such events happened in the distant past by a study of the forms in which minerals occur, and relating these to the pressures and temperatures at which their formation must have occurred.

Our knowledge of geochemistry is still very inadequate. But it is a branch of chemistry that is attracting considerable interest today, not only because it can help to satisfy man's intrinsic curiosity about his home planet, but also because it may lead to new and more accurate ways of identifying deposits of those elements needed by industrial civilization.

Some 99·6 per cent of the Earth's mass lies buried in the core and mantle, beyond our reach. The crust, from which all our exploitable minerals come, contains only 0·375 per cent of the mass. Even less is contributed by the remainder, the hydrosphere and the atmosphere, with the latter accounting for only 0·0001 per cent of the mass. Yet it is largely these two tiny portions that maintain life on the planet, although it was probably the existence of life that gave the atmosphere its present composition. In early times, it was probably quite different and, by our standards, very unpleasant. As Von R. Eshleman of Stanford University wrote: 'It is now clear that the presence of life has been, and continues to be, a controlling influence on the composition of the Earth's atmosphere. Conversely, the apparent absence of life on Venus and Mars may explain much about the nature of their atmospheres.'[3]

What are the atmospheres of our neighbouring planets, and how do they relate to our own? Space probes have made it possible to determine accurately the composition of extraterrestrial atmospheres. On both Mars and Venus, the main component of the atmosphere is carbon dioxide (0·03 per cent of Earth's atmosphere); Soviet spacecraft have shown that,

3. Von R. Eshleman, 'The atmospheres of Mars and Venus', *Scientific American*, ccxx, 3, 1969, p.78

in the case of Venus, this compound accounts for 97 per cent of the bulk of the atmosphere. On both planets, oxygen is virtually absent and very little water vapour is detectable. Is this what the Earth's atmosphere was like before life? According to Eshleman, if life ceased, the terrestrial atmosphere could become like those of Mars and Venus, with oxygen and nitrogen gradually being removed from it and tied up in inorganic compounds.

It is believed that all the water now present in oceans and atmosphere (80,000 cubic miles of sea are evaporated each year into the atmosphere and returned to ground-level as rain) must have been tied up in rocks, such as hydrous silicates, and only gradually appeared as surface water during the first thousand million years of Earth's history. From this water came atmospheric oxygen, formed by photochemical decomposition in the upper atmosphere (probably with loss of the hydrogen into outer space) and later by biological action. Oxygen now accounts for nearly 21 per cent of the Earth's atmosphere.

There are at present conflicting theories about the composition of the early atmosphere. According to the Nobel prizewinner Harold Urey, the primeval atmosphere was composed of methane, ammonia and water vapour. However, this supposition was based on the hypothesis that the whole of the Earth's formation was a cool process, which now seems to be generally disbelieved. Currently more in favour is the view that, after the planets had formed, the solar system's residual nebula, and the early planetary atmospheres were blown away by changes in the Sun. After this, any planetary atmospheres that formed did so by degassing of the planets' rocks. In this case, the Earth's primitive atmosphere is likely to have been composed of water, nitrogen and carbon dioxide, with traces of methane and ammonia, and possibly some hydrogen and carbon monoxide. What is certainly true is that the atmosphere now is quite different – and that it is life which has made it so.

## The beginnings of life

How life originated is a question that man has tried to answer for thousands of years. Early attempts – from the ancient Greeks onward – provide a fascinating insight into the way in which men's minds have worked, but contain precious little that we would call science.

The modern approach to the origin-of-life problem dates back to the 1920s. It was in 1924 that a young Soviet biochemist, Alexander Oparin, published a short pamphlet in Russian suggesting that life could have arisen from inanimate matter over a very long period, by purely natural processes. On the face of it this appears to be a plausible hypothesis, and it might seem strange that it had not been put forward earlier. The reason for this lies in the last half of the nineteenth century. The idea of spontaneous generation had been considered by naturalist Charles Darwin, but only as a conjecture. Many others considered it in the same way – as a vague possibility – but the extensive work of Louis Pasteur in defeating all existing experimental claims for spontaneous generation convinced most people that it just could not happen. This was despite the basic negative in Pasteur's approach. All he had done was show that specific claims of instances of spontaneous generation were due to faulty experimental techniques, and in 1878 he wrote: 'La generation spontanée, je la cherche sans le découvrir depuis vingt ans. Non, je ne la juge pas impossible.' ('I have looked for spontaneous generation for twenty years, and I have not found it. No, I do not judge it impossible.')[4]

Until Oparin, Pasteur's twenty years of failing to find spontaneous generation deterred others from looking for it. He not only drew attention once more to the problem, but suggested new ways of testing the hypothesis of spontaneous generation. Oparin is still active in pursuing this problem, on which he is undoubtedly one of the world's leading authorities.

Four years after Oparin published his initial remarks, the English biologist J. B. S. Haldane, unaware of the publication, wrote an essay on the spontaneous generation of life, that closely shadowed Oparin's in its concepts. Apart from a few isolated experiments, however, this new interest in the origin of life did not reach an experimental stage until after World War II. It has since grown enormously, and a coherent body of information is emerging that enables us to draw a likely scheme for the origin of life. We can never prove that such a scheme is correct, although if we discover life on other planets in earlier stages of evolution than our own, it may help to increase our certainty about the correctness of current views.

The major steps in our contemporary view on the development to life are threefold. First, it is necessary to have the small molecules on which life is based – amino acids, fats, sugars, nucleotides, and so on. Second, schemes must be devised to show how these could have polymerized to produce the giant molecules common to all life forms – proteins, carbohydrates and nucleic acids. Third, how did these macromolecules come together to form systems that could perform the functions which we recognize as characteristic of life?

To answer this question in depth, it is necessary to decide first what life is. A purely chemical account can avoid this issue to some extent.

4. J. Marquand, *Life: its nature, origins and distribution,* W. W. Norton, New York, 1968. My translation.

Given a chemical 'scenario' which leads to an organized self-reproducing system, one can say that life as we know it will follow. If one faces the issue directly, it is difficult to produce a definition of life that is neither too simple nor too complex. Even striking a happy medium between the two extremes produces a difficult definition.

For example, the noted British crystallographer and science historian J. D. Bernal defined life as 'a partial, continuous, progressive, multiform and conditionally interactive, self-realization of the potentialities of atomic electron states'.[5] A longer, but perhaps easier, definition of the function called life is offered by American chemist Melvin Calvin: 'That function in a first approximation seems to be the directed use of energy to create order from a disordered, or less ordered, environment: in biological terms, growth and differentiation . . . further, function seems to be to generate and transmit the "program" for growth and differentiation to another system, that is, reproduction. Finally, function includes changes in the "program" in response to a changing environment; the correlative biological terms would be mutation and selection.'[6]

Even if we fulfil the definitions of life in the terms of these distinguished scientists, we shall not end up with dogs and cats. What we shall end up with is a life potential that has somehow led, through biological evolution, to the dogs, the cats and us. What happened in between is the 'missing link', and to identify that is a task for future science. At present, work is still not complete on the earlier stages.

The first major experimental breakthrough in the chemical approach to the origin of life came in 1953, when Stanley Miller, then at the University of Chicago, published an account in the weekly magazine *Science* of an experiment in which he had produced amino acids and other simple 'biological' molecules under 'prebiotic' conditions.

Prebiotic conditions are those believed to have prevailed on Earth prior to the emergence of life. Miller, a pupil of Harold Urey, whose views on the Earth's early atmosphere have already been mentioned, took as his prebiotic conditions a reducing (oxygen-free) atmosphere composed of methane, ammonia and water vapour. In order to produce more complex compounds from this mixture, it was necessary to have an energy source. Miller circulated his mixture of vapours through a chamber in which there was an electric spark discharge. The compounds formed in this discharge were then swept on by the circulating gases into a water reservoir, where

5. J. D. Bernal, *The Origin of Life,* Universe Books, New York, 1967, p. xv.
6. Melvin Calvin, *Chemical Evolution,* Oxford University Press, New York, pp. 106–107.

they could be tapped off, separated and identified by chromatographic methods.

Miller's experiments were not the first in the field. In 1938 German scientists, using short wavelength ultra-violet radiation as the energy source, had converted carbon dioxide and water to formaldehyde and glyoxal, and it had been known since the 1860s, following the work of the Russian chemist A. M. Butlerov, that formaldehyde could be polymerized to sugars. In 1950, Melvin Calvin's research group at Berkeley, California, had irradiated carbon dioxide and water in the presence of ferrous ions with radiation that was intended to simulate the type of radioactivity that might have been present in the Earth's crust in prebiotic times. Formic acid and formaldehyde were formed.

Despite these early experiments, it was Miller's work that excited other chemists and gave the impetus to what is now a world-wide experimental study of prebiotic-type syntheses. The reason for the excitement was that Miller produced amino acids, the basic building blocks of proteins. Even more interesting was that the preponderance of amino acids formed were of the alpha type found in natural proteins, rather than isomeric types, such as beta-alanine, which occur only rarely in nature.

Miller's experiments also produced a host of other compounds: simple organic acids, such as formic, acetic, and propionic; urea; and, at that time considered strange as a 'biological' chemical, hydrogen cyanide (HCN). A deadly poison to current life forms, hydrogen cyanide has now achieved a pre-eminent place among the list of simple molecules that were probably involved in the emergence of biological materials. It has turned up in nearly all experiments and, during the 1960s, significant experiments showed how crucial its role may have been. In company with nearly three dozen other molecules in the last half-decade, it has also been detected in outer space.

Following Miller's first experiments, Calvin and his colleagues repeated the work with the reducing atmosphere, but used electron bombardment as their energy source, again simulating radioactivity from the young Earth's crust. They produced an even larger range of products, including adenine, one of the four base compounds found in the nucleic acids which play a key role in transmitting genetic information from one generation to another.

During the 1950s and 1960s an immense range of experiments was carried out. Each needed an energy source, and all the possible sources of energy that might have been present were used. Crustal radioactivity simulation has already been mentioned; it has been estimated that the radioactivity from the potassium isotope $^{40}$K must have been much higher in primeval times than it is now. Similarly, in a reducing atmosphere there would have been no ozone shield in the upper air to keep out ultra-violet radiation, and the Earth's surface would have received much more of this, which is of sufficient energy to cause a variety of reactions. Lightning storms could, like Miller's electrical discharges, have been potent sources of energy, as could the heat produced by volcanoes – both effects believed to have been more frequent and extensive on the early Earth than now. Last, but not least, is the possibility of energy having been derived from meteors colliding with the Earth.

First suggested by A. R. Hochstim in the early 1960s, the idea that meteoritic impacts could provide energy for chemosynthesis came from the American space programme. When the first *Mercury* space capsule re-entered the Earth's atmosphere in February 1962, radio contact was lost with astronaut John Glenn, which caused great alarm on the ground. The space capsule was, however, safe, and radio contact was soon re-established. It was not long before space scientists realized that the space capsule hurtling through the atmosphere had heated up the surrounding air by friction; this had consequently ionized the atmospheric gases, and it was this ionized layer around the capsule which had prevented transmission of radio waves.

Hochstim thought of what might happen if this enormous energy were to come from a meteorite entering the primeval atmosphere. If a spherical body tears through the Earth's atmosphere as a result of gravitational pull, a number of different regions of surrounding air can be identified. For a meteorite of 1 m radius travelling at 5 km per second (about 11,000 miles per hour), Hochstim calculated that the temperature in the stagnation region in front of the meteorite would be 7,000 K[7], and the pressure 200 times atmospheric. He also estimated that the stagnation region would hold 3 kg (6·6 lb) of material.

For a larger meteorite, the figures increase staggeringly: with a radius of 500 m and a velocity of 11 km per second, the compressed wave of the stagnation region holds 400,000 tons of material, with another 200,000 in the inner wake. The stagnation region has a temperature of 16,500 K and a pressure 1,500 times that of the atmosphere. Such conditions would produce highly reactive atomic particles. If the reaction was suddenly 'quenched' by rapid cooling and return to normal pressure, the reactive particles might form a wide variety of compounds. Many meteors im-

7. Scientists frequently measure temperatures on the absolute or Kelvin scale. One degree on this scale is the same as one degree on the Centigrade or Celsius scale, but the freezing-point of water occurs at 273.15 K, rather than zero.

pacting on the primeval Earth must have been quenched in just this way by impact with the sea.

Although large meteorites do not fall to Earth very often – one every 10,000 to 20,000 years – they probably did so at a greater rate in the Earth's early days. It has been estimated that a large meteor (500 m in radius) passing through the Earth's atmosphere now would produce 100 million kg of nitrogen monoxide (NO) and a million kg of carbon dioxide ($CO_2$). It seems likely that such large-scale chemosynthesis in the primeval atmosphere may have contributed significantly to the formation of the original organic molecules.

One of the major factors in considering likely energy inputs for primordial chemosynthesis is their deleterious effects on organic molecules. Early workers on synthetic systems found that the longer they left their apparatus running, the more biological molecules formed initially broke down as a result of the energy source acting destructively on them. Clearly, sources of energy such as meteors or lightning storms are sporadic; the energy is only available in short bursts, so the destructive effects would be minimized. But for a stock of molecules to build up, a protective mechanism must have been necessary.

The need for such protection led to the view that molecules may have concentrated in the sea to form a dilute molecular soup in which they were less likely to be exposed to the deleterious effects of radiation. Various additional suggestions have been made. For example, molecules could have been adsorbed on to moist sand or clay at seashores, where they would not only have been protected, but also concentrated. Similarly, isolated pools or lakes could have served as reservoirs for the slowly growing mass of organic molecules. If these lakes were subject to evaporation, they too could gradually have concentrated the organic matter.

Most of the experiments on organic chemical synthesis under prebiotic conditions have used 'reducing' atmospheres. Yet, as was pointed out on page 71, the Urey hypothesis of a primeval reducing atmosphere is now out of favour. Does this discredit the work that was based on it? In the words of Leslie Orgel, a leading researcher in the field, 'while the composition of the primitive atmosphere is a subject of great intrinsic interest, it doesn't have so much significance for the origins of life as some people would claim'.[8] This is because experiments conducted with different atmospheres also seem to lead to the same 'biological' type of molecules, and the continued discovery of more molecules – some, such as

8. Leslie Orgel, 'The synthesis of life molecules', in *In the beginning . . . The origins of planets and life* (ed J. P. Wild), Australian Academy of Science, Canberra, 1974, p.90

vinyl cyanide, quite complex – in outer space is changing chemists' views about the difficulty of forming the building blocks of life under 'hostile' conditions. Recently, the astronomers Chandra Wickramasinghe and Sir Fred Hoyle have suggested that outer space may have been the best environment for the formation of complex organic molecules. They claim that life itself may have originated on grains of interstellar dust and reached the earth on meteorites.

**How the little molecules grew**

The second stage in the origin of life was the formation of much larger molecules, the biological polymers. Most important biological polymers are formed by condensation reactions in which water is removed from the molecules as they link together. In very dilute solutions, the reverse is likely to happen. Small bits of polymer will tend to be hydrolyzed to their basic ingredients. In consequence, doubt has been thrown on the idea that polymers could have formed in the oceans, and one argument is that there must have been a concentration mechanism to bring molecules sufficiently close together in sufficient numbers to allow polymerization to take place. Other mechanisms have been proposed that might have helped polymerization in an unfavourably aqueous environment; most of these involve the poisonous little molecule, HCN.

In 1960 Juan Oro polymerized concentrated mixtures of HCN and ammonia and found the important biochemical adenine in the mixture. Later it was found that at high temperature this mixture could produce a number of amino acids, such as glycine, alanine, aspartic acid and serine. (The reaction did not actually form these compounds as such, but formed reactive intermediates which, on contact with water, hydrolyzed to the amino acids – and no one doubts the presence of abundant water on the early Earth.)

Miller's first projected mechanism for the formation of amino acids involved HCN as an intermediate, although as a substance that added to another molecule, rather than polymerizing with itself. But, as Calvin wrote: 'The appearance of many of the nitrogen-containing compounds in the earlier primitive atmosphere experiments is now emerging as being due to the presence of HCN.'[9]

Calvin has also shown how HCN may have formed. In a mixture of methane and ammonia, an equilibrium is set up in which these two compounds can react to form HCN and hydrogen. At normal temperatures, equilibrium favours the starting materials – the reaction needs a lot of energy to produce HCN. As the temperature rises, the situation changes, as is the case for the reaction between calcium oxide and carbon dioxide

9. Calvin, *Chemical Evolution*, p.134

(p. 40). The temperature at which this happens is readily imaginable on the early Earth and if, as seems likely, the hydrogen formed escaped into space, more HCN would have been produced as the reaction struggled to reach equilibrium.

Even if one does not accept the presence of reduced carbon in the early atmosphere, formation of HCN is not precluded. A mixture of hydrogen, nitrogen and carbon monoxide will react to produce HCN and water. Unlike the previous reaction, the equilibrium of this can be used to favour product formation if it is assumed that water was frozen out of the system – a possibility in polar regions. There is also the possibility of volcanic production of HCN. Lev Mukhin, of the U.S.S.R. Academy of Science, has calculated, on the basis of a study of volcanic areas, that a single eruption may produce in excess of 10 tonnes of HCN.

It has been known for 150 years that HCN, in aqueous solution, will form polymers. What has not been known until recently is the molecular form of these polymers. The polymer formed, with energy supplied thermally or photochemically, is always a dark, at times completely black, material. The first few stages of this polymerization have now been worked out extensively. In 1970, Monsanto chemists Clifford Matthews and Robert Moser showed how the HCN polymerizes to its tetramer (four molecules) cis-enaminonitrile. Under the influence of ultra-violet light this isomerizes and rearranges to an imidazole (the imidazole ring forms part of the side-chain of the amino acid histidine, which is essential to the catalytic activity of a number of enzymes). This imidazole can add a further molecule of HCN to form adenine. Alternatively, it can be hydrolyzed, then add another molecule of HCN to form guanine. Both adenine and guanine are essential base constituents of nucleic acids (Figure 9).

A decade before this, a Japanese chemist, S. Akabori, had proposed a mechanism for the formation of protein from HCN. According to his scheme, after polymerization of three molecules of HCN to form a trimer, hydrolysis, followed by loss of carbon dioxide, would give the amino acid derivative glycinonitrile (glycine is the simplest of the amino acids; imagine a methane molecule in which one hydrogen has been replaced by an *amino* group and a second hydrogen by a carboxylic *acid* group; in glycinonitrile, the carboxylic acid group is replaced by the nitrile group, –CN).

Glycinonitrile, according to Akabori, could polymerize to polyglycinimide, which could hydrolyze to polyglycine. If the polyglycine were then to react with aldehydes (also formed in prebiotic synthesis experiments), it could form a variegated polymer of different amino acids.

Matthews and Moser showed in 1967 that peptides could be obtained

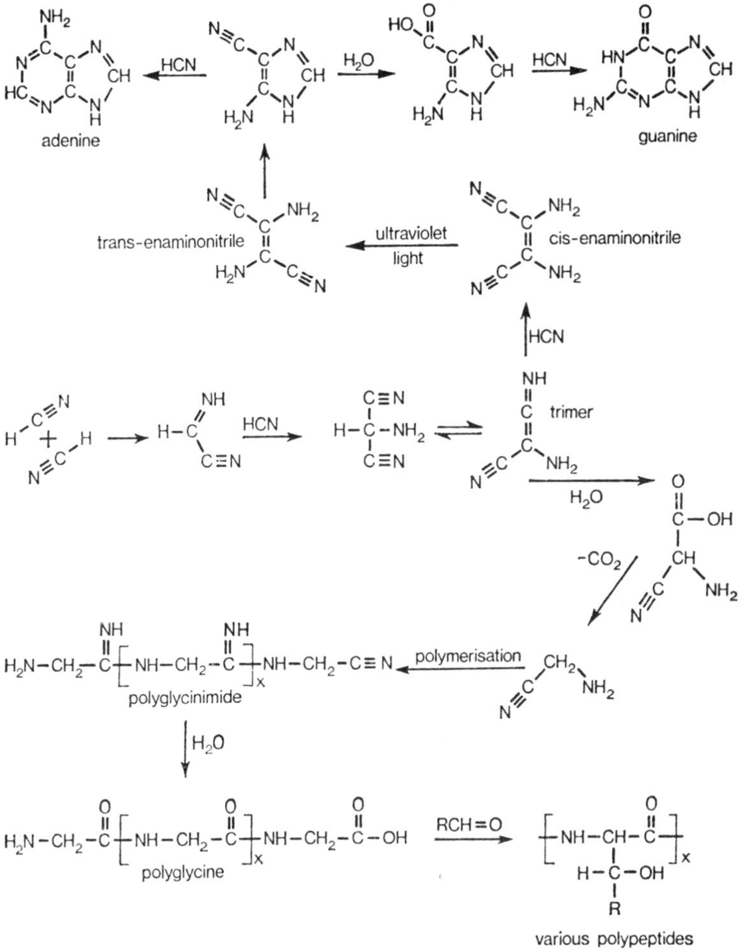

FIGURE 9 A possible scheme for the synthesis of 'biological' molecules from hydrogen cyanide (HCN).

by water treatment of the polymer formed by concentrated mixtures of HCN and ammonia. However, they attributed to ammonia only the role of a catalyst, and produced a scheme to show that HCN by itself could produce a polymer which would turn into a protein when treated with water. From their reactions they identified fourteen of the common amino acids. The other half-dozen normally present in proteins, with side-chains that cannot be formed from these few components, could have arisen from reaction with other simple molecules likely to be available on the primeval Earth, such as acetylene and hydrogen sulphide.

While work on the polymerization of HCN is leading to a possible origin for protein, it is not the only procedure being investigated. This is fortunate, for it is a scheme that skirts certain problems. In the Miller-type experiments, amino acids as well as HCN were found, and Matthews and Moser have suggested that these were obtained from the breakdown of HCN polymers. At some time in the evolution towards life, certain proteins – notably those with catalytic properties, the protoenzymes – must have become favoured molecules. It would no longer have been possible to rely on the proteins formed by chance from HCN and then reacted with other materials to give the different side-chains. It is necessary for us to find a mechanism that can polymerize amino acids into proteins in a regulated way.

One of the methods that has been investigated by Calvin and his co-workers again involves the ubiquitous HCN molecule, but in a different role. From HCN and ammonia, the compound cyanamide can be formed. Although the formula for this is generally written $H_2N—CN$ (a *cyan*ide group attached to an *amide* group), each nitrogen is actually bonded equally strongly to the carbon atom (delocalized molecular orbitals) so that the central structure is more like —N = C = N—. Cyanamide can polymerize to dicyandiamide or, alternatively, can react with another HCN molecule to form dicyanamide, in which the amide has two cyanide groups attached to it. All these molecules contain the —N = C = N— structure.

Calvin noted that this structure also occurs in a class of compounds called carbodiimides, which have been used for some years on a laboratory scale in the chemical synthesis of peptides from amino acids. These compounds remove the water formed during the type of condensation polymerization that is involved. The actual parent compound, carbodiimide, is none other than the H—N = C = N—H form of cyanamide. When the elements of water add to this, they form the simple compound urea, another of the substances found among the products of the early Miller-type synthesis.

In 1966, Calvin and his colleagues dripped a solution of the simple amino acid glycine into a slightly acid solution of dicyandiamide. Analysis of the mixture disclosed the peptide tetraglycine (formed by the condensation of four glycine molecules), and its lower molecular weight analogues made, respectively, from three and two glycine molecules.

It is not only the condensation of amino acid molecules that can be assisted by carbodiimides. Carboxylic acids, in the presence of alcohols, are converted to esters, and it is worth pointing out that fats which, although not macromolecules, are still an important class of biological molecules, are esters of carboxylic acids and the small molecule glycerol. Similarly, phosphoric acid or its derivatives can be condensed to pyrophosphates which may, in turn, serve as dehydrating agents, and are now the principal class of agents used in living substances for this purpose.

This is not the only way in which pyrophosphates may have arisen. In 1964 Miller showed that if the mineral calcium hydroxyapatite (in which phosphate is the major anion) was treated with potassium cyanate (an inorganic substance which contains a modified cyanide grouping), pyrophosphate was released on heating. However, this is a controversial area and demonstrates some of the uneasiness that can arise when scientists borrow information from disciplines other than their own; in this case, chemists are borrowing from mineralogists. Often, a more experienced person, in the field borrowed from, may raise objections. Here is the late J. D. Bernal, mineralogist and crystallographer, on the subject of phosphate:

> Phosphorus is not a very abundant element on the crust of the Earth and its liberation from the rocks, on the one hand, and its fixation in organisms, on the other, determine the total rate of life activity on this globe.
>
> Much may depend on the various forms in which the phosphate ions were first liberated from the basic mineral, apatite, $Ca_5(PO_4)_3F$, and even on the detailed texture of this mineral, because it would appear that the activity of an apatite as it yields up the phosphates, is itself a function of the imperfection of its individual crystals. The various dislocations which may occur in apatite crystals may be actually the way in which the phosphate radicals appear in pairs as metaphosphates and thus enter for the first time into biological combinations, because the orthophosphates are at once very little soluble and very stable compounds.[10]

Metaphosphoric acid, $HPO_3$, can formally be converted to ordinary or orthophosphoric acid by addition of one molecule of water. Similarly, in the formal sense, one molecule of water shared between two molecules of metaphosphoric acid will give pyrophosphoric acid. In fact, metaphosphoric acid is more complex than it appears, since it polymerizes with itself to form chains and rings of metaphosphoric acid molecules. On prolonged contact with water, the viscous polymeric liquid gradually hydrolyzes to orthophosphoric acid. Chemists seem now to have centred their interest on the organic phosphate esters as dehydrating agents. It has

10. Bernal, *The Origin of Life*, p.50

been suggested, however, that the use of phosphate esters is unrealistic, as it is unlikely they would have formed on the primitive Earth.

A different approach has been taken by Leslie Orgel and G. J. Handschuh of the Salk Institute. They reported in 1973 that struvite, a mineral made from magnesium ammonium phosphate, phosphorylates nucleotides more easily than hydroxyapatite does. In addition, magnesium ammonium phosphate precipitates readily from solutions; therefore, in the slowly evaporating rock pools often favoured by theorists as sites for the origin of life, struvite may have been more readily available than hydroxyapatite.

Development of the ideas of using the carbodiimide linkage either to condense, for example, amino acids into peptides in dilute solution, or to produce the type of phosphates that will in turn perform this function, has overcome, to some extent, criticism of the 'primeval broth' idea. By such mechanisms it is possible that condensation polymerization could have taken place against the concentration gradient leading toward hydrolysis.

In this brief account, simplification has produced what may appear to be a puzzle. Why should active molecules such as carbodiimide react with water released during a condensation polymerization, but not react with the water – and a much greater amount of it – in which they are dissolved? The answer is that they do not react with water produced by condensation polymerization. They add the elements of water during the condensation, a little at a time. Although they will react with water, they react preferentially with the carboxylic acid end of an amino acid, to form a substituted amino acid. The new compound formed is reactive toward the amino end of another amino acid and, in linking with this, it expels the carbodiimide which, in these two stages, has gained the elements of water. There is no actual reaction with water molecules, but a multi-stage reaction which gives the overall appearance of reaction with water.

Having given the arguments critical of the 'primeval broth' hypothesis and shown a way around them, it would be unfair just to stop. We are not trying to produce a single picture of the origin of life. At the stage of big molecule formation there were likely to have been multiple origins rather than merely an origin, just as it is likely that the original chemosyntheses of starting materials used more than one energy source.

Condensation polymerizations can be performed by supplying thermal energy which drives off water as steam and leaves a polymer behind. One of the alternatives proposed to the oceanic origin of life is that it might have occurred in isolated pools, where alternate heating and cooling could concentrate the organic ingredients so that they would polymerize. Subsequent cooling and rainfall, reconstituting the pool, would then have allowed them to interact with one another.

A leading proponent of this theory is Dr Sidney Fox, Director of the Institute of Molecular Evolution at the University of Miami. Fox's interest in the origin-of-life problem goes back a number of years, and covers the formation of the original small molecules as well as the polymers. In the early 1960s, together with Dr Kaoru Harada, he synthesized a large number of amino acids of the type used in protein building by passing a mixed vapour of water, ammonia and methane at high temperatures through a tube packed with particles of silica or quartz, thus showing how high temperature might have provided energy for chemosynthesis, and also indicating the possible involvement of the mineral surfaces as catalysts.

For more than a decade Fox and his co-workers have been studying the properties of proteinoids, materials resembling proteins which they have synthesized under the intermittent hot/cold conditions which they believe formed the seedbeds of life. In the late 1950s, Fox found that if a mixture of amino acids were heated to 150 °C and then plunged into water, insoluble proteinoids were formed which had the same type of chemical linkage as real proteins.

To achieve these results. Fox had to use very high concentrations of aspartic and glutamic acids or lysine. In a typical experiment, a mixture composed of 33 per cent aspartic acid, 17 per cent glutamic acid, and 50 per cent mixture of equal parts of seventeen other amino acids, was heated together at 100° for 150 hours. Alternatively, such a mixture could be heated at less than 100° in polyphosphoric acid solution to give proteinoids.

Fox and his colleagues now claim to be able to produce proteinoids from mixtures of amino acids in which the aspartic or glutamic acids or lysine are at no higher concentration than the other amino acids. This is an important step forward, since a criticism of Fox's early work was that it was unlikely that there would be such large excesses of these three particular amino acids.

Molecular weights of the proteinoids are between about 3,000 and 11,000. The molecular weight of glycine is about 75, that of tryptophan (the heaviest amino acid) just over 200, with many of the others falling in the range of 125 to 175. Fox's proteinoids probably contain somewhere between 20 and 75 amino acid residues, which is small by the standards of present-day proteins.

One of the most interesting results is the selectivity of the proteinoids. Superficially, one might imagine that heating a random mixture of eighteen amino acids would produce random proteinoids. But in the typical early experiment mentioned above, although 33 per cent of the starting material was aspartic acid, this amino acid accounted for 51 per cent of the proteinoid material. Other amino acids which were all present in about 3 per cent concentration in the original mixture were incorporated into proteinoid in amounts varying from nearly 6 per cent for phenylalanine to barely half a per cent for threonine.

This selectivity is important, as is the discovery that if one starts with all the amino acids present as the same optical isomer, the type found in nature, for example, the proteinoids are optically active, although unreacted amino acids are found in the reaction product as racemates of both isomers. An interesting sideline of this later work is that the proteinoids have nutritive value, and are presently being investigated from this viewpoint, presumably as a potential food additive to improve natural protein quality.

The proteins synthesized by living organisms today have specific structures, and it is clear that their functions – as enzymes, connective tissue, and so on – depend closely on their structures.

Early in the search for clues to the chemical origins of life, it was believed that amino acids would have polymerized randomly to a great variety of proteins, only a tiny fraction of which would have been of any use. This implies a great element of luck in the formation of a significant protein configuration, and also suggests that all non-viable proteins would have been destroyed by natural processes, either by weathering or by being eaten. However, recent years have shown that nothing in chemistry is as random as it could be, and it may be, as Bernal said, that life is a function of the electronic configuration of atoms. A logical pattern might be built up from the atoms themselves, with no element of luck involved.

Fox's proteinoids are an important part of the evidence for this. On polymerizing, a random mixture of amino acids tends to reduce its own randomness. This is not so surprising, for when we talk about a random mixture of amino acids polymerizing, we tend to conceptualize 'amino acids'. In reality, of course, each amino acid is a molecule with definite molecular characteristics and a specific shape, or number of preferred shapes. The chemistry of polymerization is thus three-dimensional. One must take into account not only the basic polymer chain which forms from the amino acid backbone, but the effects of the different side-chains (the groups of atoms which stick out from the backbone).

For example, if an amino acid has a bulky side-chain, like tryptophan, is it more difficult for its backbone to fit into the right place next to the chain for condensation to take place? Or, since long chain carbon molecules with no other substituents but hydrogen have an affinity for one another, will an amino acid like leucine be selected preferentially by the growing end of a protein chain in which the preceding addition to the chain was a similar amino acid – leucine, isoleucine, or valine – over one of a quite different character, such as aspartic acid?

Once the chain begins to lengthen, it takes up a three-dimensional configuration of its own that may bring disparate parts of the molecule together. Therefore we must not exclude the possibility that the character of an amino acid at an early position in the chain may affect the choice of additions much further away than would appear possible if we looked on the molecule as a long chain spread out in a nearly straight line. Much of this is conjectural, but testable. One of Calvin's colleagues, Gary Steinman, has done some work on the problem. He attached glycine to a polymeric (non-protein) material and reacted this with amino acids to form dipeptides. In each experiment, equivalent amounts of the amino acids were used, but the yields from the reactions differed widely. Steinman then compared the yields of his dipeptides with the frequencies of dipeptide occurrence in all natural proteins for which the amino acid sequence was known. Although the figures differed, the rank order in each case was the same: the dipeptides that had formed in the poorest yields were also those found least often in known protein structures.

Proteins are manufactured in animals and plants according to instructions laid down in the nucleic acid code. When did nucleic acids take over the control of protein sequences? Did the interaction take place after the origin of useful proteins, so that at least some of today's proteins are descendants of protoprotein that arose according to strict chemical laws, and not the strict instructions of the genetic code? Was there some sort of intermediate stage? Dr Charles Carter, of the University of North Carolina, has carried out some studies which show that ribonucleic acids and proteins can form complexes which fit neatly together in stable helical configurations. He has suggested that either component of such a complex might have acted as a template for synthesizing the other component from suitably activated starting materials in the prebiotic soup.

Another major question relates to the optical isomerism of amino acids. All the primeval broth experiments, following Miller, produced predominantly the alpha amino acids found in nature, but they were produced as racemates – equal amounts of both optical isomers. Yet all life today is based on only a single type of isomer.

Just as a growing polymer may affect its own continued growth, randomness can also give way to selectivity of isomeric forms. In 1954 E. Havinga showed how such selectivity may arise in a series of compounds called quaternary amines. These occur in two isomeric forms which rapidly interconvert with one another in solution. When he crystallized solutions of the amines slowly, Havinga found that all the solid which precipitated did so in only one of the two possible isomeric forms. Because of its three-dimensional structure, whichever isomer happened to form the first crystal nucleus encouraged all the other molecules to adopt its isomeric form as they deposited on it. A growing protein can influence its further growth in a similar way. Fox's proteinoids from a single isomeric type gave proteins of the same isomeric type, although the reaction product contained unreacted acids in both forms.

An 'optically pure' amino acid polymer will take up a helical configuration, like a spring, with hydrogen bonds, which are a particular type of electron-excess bond, forming between any one turn of the spiral and the next turns above and below, thus helping to stabilize the structure. The only difference between homopolymers of the different optical isomers is that the helix turns in different directions. In the naturally occurring protein it is called the alpha helix. If an optically impure polymer is formed, it will tend to twist one way and then another, according to the configuration of the last few monomers added to the chain. It will not produce a structure stabilized by hydrogen bonding, and is therefore more tedious to synthesize. We tend to think that everything moves toward increasing chaos, but here we can see that simple chemical considerations may lead to increasing order as a more natural process.

It is possible that selection between left- and right-handedness takes place at a more fundamental level. The physicist's universe is asymmetric: for example, electrons emitted during beta-decay of radioactive nuclei are predominantly 'left-handed'. Experiments carried out in recent years, notably by A. S. Garay, working at the Hungarian Academy of Sciences in Szeged (from which he fled to the West in 1975), have shown that one optical isomer of the amino acid tyrosine decomposes faster than its mirror image in the presence of beta radiation.

It has always been assumed by chemists that mirror-image molecules are identical in all aspects of their behaviour other than their effect on plane polarized light. It seems now that this is not so. Einstein's relativity theory showed that Newton's laws of physics were inadequate to describe the Universe at large. Nevertheless, the great majority of us still live our lives according to these laws. Similarly, for most practical purposes, chemists will continue to look on optical isomers as having identical properties. But, at the fundamental level of why a single optical form predominates throughout nature, it may be the particle physicists who have the final word.

The hypothesis that biochemical biases have their origins in particle asymmetry has one interesting consequence. It has been assumed that, elsewhere in the universe, life will have developed (if at all) on similar chemical lines. The protein structures would be different; the genetic code would be undecipherable by our metabolism; but the feeling has been that the basic building blocks would be broadly comparable. The discovery of abiologically formed organic molecules, such as amino acids, in meteorites has strengthened the belief that chemical constraints will have led to the choice of similar building blocks in other places.

Writing along these lines recently, the German Nobel prizewinner Manfred Eigen said 'individual compositions and arrangements, especially the molecular code and mechanism, will little resemble the systems known to us, although they will probably be built on similar principles. In particular, there should be a 50 per cent chance of finding a different chirality (handedness) in optically active structures.'[11] If, however, it is the asymmetry of fundamental physics that underlies the asymmetry of life then, provided we are in a 'matter' (as opposed to an 'antimatter') universe, the handedness ought to be the same, wherever we find life.

**Toward life's master molecules**

The structure of proteins is, in a sense, simple: amino acids have a common backbone which polymerizes, and this backbone is not a complex structure. The nucleic acids are different, for they are made up of three components: the nucleic acid bases, five in number, the sugar ribose, or its analogue-2-deoxyribose, and a phosphate group. If one takes a hydrogen from orthophosphoric acid and a hydroxyl group from a particular point in the ribose or deoxyribose molecule and links the two residues, then removes the elements of water from one of the bases and another part of the ribose molecule, the result is a nucleotide, the basic building block of the nucleic acids. The polymer itself is built up from such tripartite groups, by loss of the elements of yet another water molecule from the phosphate part of one unit and the sugar part of the next.

Before considering the polymerization of the nucleic acids and looking for directedness,[12] it is important to backtrack a little, for there are problems in the formation of the tripartite monomer unit. Why, for example, are the molecules of water lost in such a way that the phosphate always links on to the sugar in one place, and the base in another place, but it is always the same place that either of these two components is linked to the central sugar?

11. Manfred Eigen and R. Winkler-Oswatitsch, 'The game of evolution', *Interdisciplinary Science Reviews*, i, 1976, p.19.
12. Directedness is the property which makes the structure of the large molecules inherent in the building blocks, although not obviously so.

The five bases commonly found in nucleic acids are called cytosine, thymine, uracil, adenine and guanine. The last two can be produced by polymerization of HCN, as already mentioned. There is now even a Japanese manufacturing process for producing adenine in tonnage quantities from cyanide and ammonia. In the early 1960s, Oro and his colleagues not only synthesized adenine from ammonium cyanide, but also obtained 2-deoxyribose from formaldehyde and acetaldehyde in aqueous salt solutions. Subsequently Cyril Ponnamperuna synthesized both ribose and deoxyribose by irradiating dilute formaldehyde solutions with ultraviolet or $\gamma$-rays.

There has been much concentration on adenine and guanine, the most complex of the five bases used in nucleic acid synthesis. The other three have been rather neglected and do not seem to have appeared as products in many prebiotic synthesis experiments. Uracil has been synthesized from urea and maleic acid, both simple compounds that have appeared as the products of prebiotic synthesis experiments, in the presence of polyphosphoric acid. In 1968 cytosine was synthesized from cyanoacetylene and cyanate. Thymine, the remaining nucleic acid base, could similarly be prepared from urea and the 2-methyl homologue of maleic acid.

In 1964 G. Schramm reacted low concentrations of adenine and ribose in a non-aqueous solvent, and obtained a 20 per cent yield of adenosine, the compound in which adenine and ribose are coupled (as in nature). Ribose has six hydroxyl groups on it which could react with adenine, so a 20 per cent yield of the 'natural' material implies chemical selectivity. On treating nucleosides (the compounds made from the sugar and nucleic acid base) with polyphosphates, Schramm obtained polynucleotides (smaller polymers than the natural nucleic acids, but having the same structure). As the conditions used by Schramm are similar to those chosen by Fox for his work with amino acids, the intermittent heating/cooling in isolated areas can account for the rise of nucleic acids as well as of proteins.

One of the great triumphs of modern science has been the discovery of the fundamental nature of the nucleic acids, particularly the discovery – described in more detail in Chapter 4 – by Francis Crick and James Watson of the double helix, the ordered conformation by which two complementary chains of nucleic acid form helices that are stabilized in an intertwined double helix by hydrogen bonds between bases on both helices. In deoxyribonucleic acids (those in which the sugar moiety is 2-deoxyribose), adenine hydrogen bonds to thymine and cytosine to guanine; in ribonucleic acids, uracil substitutes for thymine.

In searching for the ways in which order might have developed from disorder during the pre-evolution of life, it was only reasonable that attempts should be made to see what effect one strand of a double helix has on the synthesis of its complement. It was found that in the presence of polyadenylic acid, a homopolymeric nucleic acid in which the base is adenine, the polymerization of uridine monophosphate (uridine is the condensation product of uracil and a ribose sugar) occurred at a greater rate than in the absence of the adenylic polymer. Similarly, Calvin found that if one linked together hexathymidines to the dodecamers (two six-chain units forming one twelve-chain unit), using carbodiimides as condensing agents, the rate of reaction was increased tenfold when polyadenylic acid was present. Clearly, the existing polymer, by hydrogen bonding to its complementary fragments, brings their reactive ends more accurately into proximity for reaction than if there is nothing to hold them in place.

The idea of molecular directedness has brought about a great change in thinking about the origin-of-life problem in the quarter century during which experimental studies have been under way. From setting out to prove that random chemistry could, on a large time scale, produce the requisites for ordered life to form, we have actually found order at a much earlier stage. This means that hypotheses can involve a shorter time scale, which is proving to be necessary as older and older fossil remnants shorten the speculated period between formation of the earth and the first true living organisms.

As Calvin wrote: 'The high efficiency in all of these stages of the polymer formation, the replication system and the coupling of the replication system will depend upon ordered chemistry. This chemistry is different from what we are accustomed to thinking of when we put reactive reagents together in a flask, all of them randomly distributed as individual molecules.'[13]

Even with the order inherent in the molecules, they still need a container that is smaller than, for example, a rock pool. At some time discrete systems had to form, protected from the environment as a whole by a skin – the beginnings of present-day cells.

Polymers of a biological type could have been synthesized on the primeval Earth from the supply of simple organic compounds formed by the interaction between energy and the constituents of the early atmosphere. Once the synthesis took place, it created another difficulty which had to be solved before life could originate. The polymers gradually used up the supply of simple building blocks; if they developed simple catalytic properties, such as that shown by polyadenylic acid in the polymerization of polythymidine, they could have removed them much faster than new

13. Calvin, *Chemical Evolution*, pp.180–2.

material was formed. Gradually the supply of readily available 'food' for polymer construction would have been depleted. Some development was needed to protect polymers against breakdown and encourage the concentration of raw materials from the surrounding solution. The answer was the development of an enclosed organic unit, a protocell.

All living organisms today have cells; the simplest organisms are single-celled creatures. The contents of the cell are bounded by a membrane which can take in useful matter from the environment and eject waste products. Just because single-celled organisms now exist, it does not mean that these are also the oldest living organisms. The minute bacteria of today are highly sophisticated, and there is reason to believe that they have evolved a long way from the original life forms, just as multi-celled organisms have done.

It has been suggested, to get away from too simple an analogy which may mislead us, that we consider the possibility that the first living systems were not enclosed by cell-like membranes but, perhaps, by the physical boundaries of a rock pool; that a metabolizing system may have extended over an area of an acre. Although this may be true, cells did arise and, in order to survive, life forms needed cells. What, for example, would happen to a living pool if it were dried up by volcanic heat?

Consequently, much research effort has gone into simple cell-like systems and to showing how boundary membranes could arise under primitive earth conditions. A relevant aspect of Fox's work on thermal proteinoids is that, when his polymerized mixtures are poured into water, they separate into discrete units (microspheres) which appear, superficially, like cells. They are similar in size, shape, and tendency to agglomerate to some coccoid bacteria. They can also grow and proliferate by producing buds which eventually break off to form new microspheres.

For many years, the Soviet scientist Oparin has experimented with another type of microsphere, the coacervate droplet. These discrete droplets form automatically in solutions of different polymers, such as gum arabic (a carbohydrate polymer) and gelatine (a protein). This work has been attacked because it uses present-day biological polymers, but, despite this criticism, coacervate droplets are useful models for showing how a membrane may aid in the survival of its contents.

For example, Oparin has produced coacervate droplets which incorporate an enzyme capable of polymerizing adenine to polyadenylic acid. Into the solution in which such droplets were suspended was placed a source of adenine, adenosine diphosphate. The coacervate droplets grew visibly as polyadenylic acid formed inside them and was unable to diffuse back out of the droplets' boundaries. A by-product, inorganic phosphate ions, appeared in the surrounding medium. A simple catalytic system protected by a membrane was thus able to extract material from solution, use it, and throw away the waste products.

Fox has shown similar simple metabolism in specially prepared proteinoid microspheres. This might be assumed to weaken the argument in favour of Oparin's coacervate droplets having played a role in chemical evolution. However, experiments in which mononucleotides were polymerized to polynucleic acids in the presence of polypeptide (small protein) have shown that coacervate droplets can arise naturally – just as naturally as Fox's microspheres. Again it is important not to place too much faith in any single answer to the origin-of-life problem. In the time that life has existed, there has been plenty of opportunity for the clues to its origins to be destroyed, so we can never be sure that there was a single origin.

There are other ways in which protocells might have arisen. For thirty years Bernal advocated consideration of mineral surfaces as areas where organic molecules might have concentrated by adsorption. Seashores would be a good example. Seafoam on the surface of the oceans has a higher concentration of organic material than the water below. Therefore, one might assume that at the seashores, where molecules would be more abundant through the presence of foam and minerals, protocells might have formed. Imagine a small patch of water-insoluble, nearly straight-chain polymer floating on the sea, like a patch of oil. It will move with the water surface, following its contour. If the water is whipped up by the wind, foaming as it crashes on to the beach, the polymer patch will be converted into spheres. Some of these will have air inside (as Figure 10 shows), but others will have water inside them, and whatever that water contains in solution or suspension. Here again is the possibility of a protocell, although in this case the contents are dictated by chance.

## Assemble in an orderly fashion

There is another aspect of macromolecules that fights against chance: self-assembly. It can be illustrated by a modern-day example. Tobacco mosaic virus (TMV), which causes disease in tomato plants, is made up of nucleic acid and a protective protein sheath. These can be dissociated. When the components are mixed together again, they reassemble automatically, yet when separate they are only a pair of macromolecules. As Calvin says: 'The construction of what is marginally, if not unequivocally, a living organism, is self-contained within the structures of the polymers of which it is made.'[13]

More and more, as natural polymeric systems are studied, it is realized

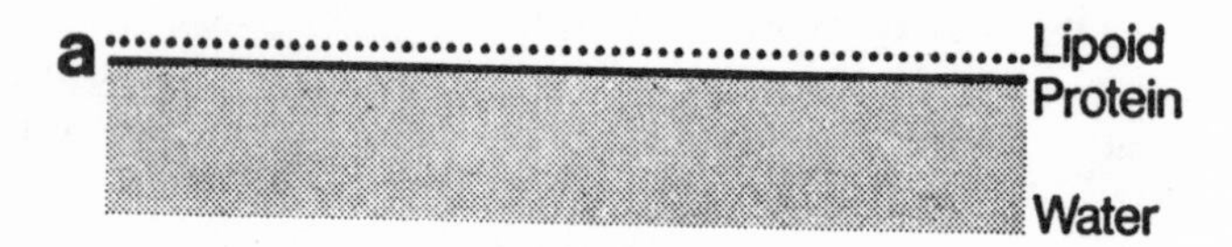


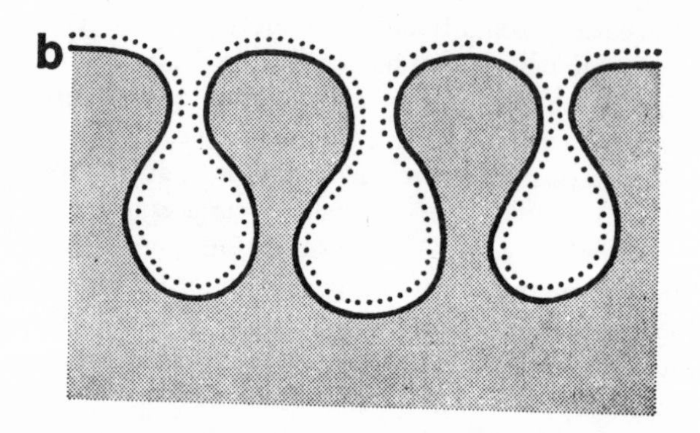


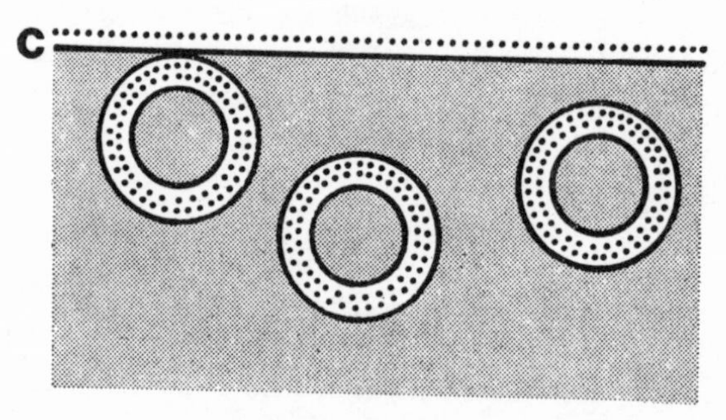


FIGURE 10 Polymeric organic molecules floating on a water surface in primeval times could have formed droplets by the action of wind and waves. Some of these droplets would have been air-filled (bubbles), others would have been filled with the water and any dissolved materials. If the skin of polymer protected the contents by forming such droplets the result may have been a prototype of modern biological cells.

that their overall structure is only a reflection of their fundamental structure. We have not yet reached the stage of predicting overall properties from the 'electron states', to use Bernal's terminology, but we are getting there.

Since the study of self-assembly is still in its infancy, research is mainly directed toward studying the incidence of self-assemblies of living systems of today. This is a very valuable study – it was, for example, disassembling TMV that made it possible to show that the infective part of the virus was the nucleic acid, and that the protein part did not contribute to the virulence – but it has meant that studies on possible self-assembly of prebiotically synthesized polymers are not yet under way.

Some of the experiments with modern materials do offer guidelines as to how primitive assemblies may have taken place, and such assemblies must have played an important part in concentrating molecules. An enzyme called transacetylase, for example, is composed of thirty-two identical protein units. The enzyme can be split into individual protein molecules, and a mixture of these will again combine into colonies of thirty-two. It is only owing to the sophistication of modern separation techniques that we know that transacetylase is an enzyme at all. In living systems, enzymes often occur in linked groups, forming a metabolic pathway in which a single molecule is modified a number of times, each modification being mediated by a single enzyme. This is rather like the game in which one word is converted into another in a specified number of steps by changing only one letter at a time, with each intermediate also forming a word (e.g., POST → P*A*ST → *L*AST → LAS*H* → LA*T*H). Transacetylase is part of such a system of three enzymes; at one time it was believed that these three were a single enzyme. They were even given a single name, pyruvic acid oxidase. If the three enzymes are split up, they will recombine into the multiple system, as fully active as it was before it was taken apart.

The reasons for this self-assembly are not known in detail, although they can be guessed at. Just as polyadenylic acid assembles to form a double helix with a polymer of one of its complementing bases, thus establishing a system stabilized by hydrogen bonding and the good physical fitting together of the components, so these more complex ordered systems must fit together according to similar rules.

If this is applied to primitive synthesis, how accurate must such fitting together be? Although the formation of proteins is not a completely random process, as Fox's experiments showed, different proteins must have formed. What would their affinities have been? Modern research on this aspect indicates that the requisites for a good fit may be very stringent – not a positive contribution to origin-of-life studies.

Molecules of the same protein from different species, in which the differences are likely to be quite small, have been shown to be able to tell each other apart. Haemocyanin, which in molluscs and arthropods plays the part played in man by haemoglobin, consists of assemblies of between 100 and 1,000 identical units. In Chicago in 1966, H. Fernandez-Moran prepared pure samples of the haemocyanins from a species of snail and a crab. Each was found to have its own specific shape. When Fernandez-Moran split the two snail haemocyanins into their single protein subunits and mixed the two, he found that they recombined selectively.

Once more J. D. Bernal has cut into the problem with an insight from his own field of study. Segregation and concentration of molecules is not a problem, he claimed; witness the segregated nature of minerals in the crust, with like and like joined together in the same minerals. This is perhaps an over-simplification of the issue, but it does bring home the fact that, at this stage of studies in the field, one cannot expect an easy answer to how life originated. The problems of sorting out self-assembly on a less organized scale remain, but there seems no doubt that they will be solved with time and patience; it must be remembered that on the young Earth there was lots of time in which they could be solved, although not as much as was once thought.

## When and how long?

According to current views, our planet was formed about 4·6 aeons (an aeon is $10^9$ years) ago. There are remnants of protozoa (single-celled organism) which are just over three aeons old, while remains of multi-celled organisms dating back 2–2½ aeons have been found. There is some doubt about the authenticity of such early fossils. During the early 1960s, great excitement was caused by studies of materials found in a particular class of meteorites called carbonaceous chondrites, which contain higher percentages of carbon than other meteorites. Analysis of these showed the presence of a number of biological materials, such as porphyrins. It was further claimed that the chondrites contained structures resembling living organisms. The structures were mineral in nature, but it was argued that they could be the mineralized fossils of living organisms. Detective work by a number of scientists, notably George Mueller, now a professor at Miami's Institute of Molecular Evolution, showed that such structures may arise in minerals in a number of ways that do not involve the intermediacy of living organisms. It has also been shown that all the organic molecules found in the meteorites can arise abiologically, as in the primitive Earth syntheses.

Having applied such rigorous testing to the carbonaceous chondrites, one must be equally severe with 'mineralized fossils' found on Earth. The whole of the Earth is contaminated with biological materials, so the presence of biological molecules cannot be used as evidence for early life forms. Most of the evidence must come from formations in rocks that appear to have organic structure. One piece of evidence for early life forms is a fossil found in rocks known as Gunflint Chert, which occur in Michigan, and are probably about 1·7–1·9 aeons old. Microfossils discovered in this Chert in 1965 were found to look like a modern organism with the almost unpronounceable name of Kakabekia umbellata. This rather rare organism was found at the base of Harlech Castle in Wales, and had the odd characteristic of requiring an environment of 30 per cent ammonia for growth. This, by the way, is why it was found at the base of a medieval castle, where the urine of generations had produced a urea-enriched environment just right for the growth of an ammonia-needing organism. But the real significance of this discovery is that if the Chert microfossil was, during its lifetime millions of years ago, an organism like Kakabekia, it strengthens the claim that the atmosphere of the primitive Earth was loaded with ammonia.

Kakabekia, or its prototype, is a youngster compared with the presumed fossil remains in Fig Tree Chert from South Africa. Dated at 3·1 aeons old, this contains objects which appear to be bacterial fossils and which, from the structure of sections of the rock, must have been there when the rock was formed. It has to be admitted, however, that these structures might be merely physicochemical curiosities, and not biological remains.

There are immense difficulties in tracing back organisms, for the organic materials of which they are composed are in many cases unstable. During the 1950s Philip Abelson analysed fossil organisms up to 300 million years old for traces of organic material, and found minute quantities of amino acids in fossil bones. It was possible to show that some amino acids were far more prevalent than others, owing to differences in stability. Much of this work could not have been done without the aid of paper chromatography, used in the separation and identification of the vanishingly small quantities of material involved.

Although we can assume that amino acids from fossil bone were produced by life processes, the presence of amino acids in ancient rocks cannot be used for dating the origin of life, for we have seen how readily amino acids can form abiologically. There are, however, molecules which, it is believed, can be used as biological markers. These are mostly hydrocarbons, which are exceptionally stable molecules, and are sought in conditions where they will have been protected from the deleterious effects of atmospheric oxygen. Hydrocarbons contain only carbon and hydrogen and can be produced, for example, by the removal of carbon dioxide from the fatty acids found in biological lipids.

Despite the great variety of metabolic processes in living organisms, there are certain features common to biological materials that one would not expect to find in abiologically formed materials of a similar nature. For example, two important pathways for the biosynthesis of complex molecules are the acetate and the isoprenoid pathways. Fatty acids are built up from two carbon acetate units, thus giving predominance to even-numbered carbon chains for fatty acids and odd-numbered chains for the hydrocarbons derived by decarboxylation. Abiological synthesis of hydrocarbons would not be expected to give such a high ratio of odd:even chains as would be found in material that had been biologically formed.

The five carbon isoprene unit is used by a variety of plants in forming such polymers as natural rubber. Its major characteristic is that the backbone of the polymer chain forms from four of these carbons, the fifth emerging as a side-chain at definite repeating intervals. Adding up all the possible hydrocarbons that can be formed from as few as nineteen carbon atoms and forty hydrogens, taking into account all possible types of branching of the chain, there are found to be about 100,000 isomers. Consequently, even a small predominance of one particular isomer for which a biosynthetic pathway is deducible can be used as evidence for the presence of life at the time the molecule was formed.

Ancient sediments, dated by isotopic techniques, have been analysed for organic content by the combined gas chromatography-mass spectrometry technique described in Chapter 2. Results so far have been promising in showing that biological markers do leave an indelible mark on history. Most of the studies to date have been done with sediments which are known to have formed much later than the origin of life, but this is a necessary study, since it enables critical evaluation of the methods used – a prerequisite for drawing conclusions from the older sediments.

In a number of such sediments, relatively high concentrations of the branched hydrocarbons phytane and pristane have been found. The phytyl group occurs as a side-chain attached to the central porphyrin ring of the important plant pigment chlorophyll a. The porphyrin ring has been found in a number of ancient crude oils, but as Fox showed in 1968 that porphyrins can form abiologically by thermal synthesis, this complex molecule cannot be accepted as a biological marker. On the other hand, phytane and its degradation product, pristane, are both isoprenoid molecules, and can be assigned a biological origin on at least statistical grounds.

The farther back one goes, the more diffuse the pattern of found compounds becomes, but the Fig Tree Chert still has an appreciable quantity of isoprenoids in its hydrocarbon fraction. Taken together with the tentative microscopical evidence of bacterial fossils, it seems that some form of monocellular life must have existed on the Earth at least 3·1 aeons ago. Another possible test for ancient life is the ratio of the $^{13}C$ and $^{12}C$ isotopes in carbon-containing rocks. The ratio of these two isotopes is altered during photosynthesis – a life process. In 1972, scientists from the University of California at Los Angeles and NASA's Exobiology Division found a discontinuity in the $^{13}C$:$^{12}C$ ratio in rocks formed about 3·3 aeons ago, which suggests that this may be the date of life's origin on Earth.

The main aim of organic geochemistry, as the subject is called, must now be to push the frontiers back even further, to try to find organic materials that show a gradation in biological content which can be linked to their age. In this way it may be possible to plot a time scale for the transition from preformed organic molecules to molecules synthesized biologically. This is a challenge so large that it could keep chemists occupied for the rest of this century, or even longer. In the meantime, it is a study that will be greatly helped, and stimulated, by the study of extraterrestrial rock samples.

In the rocks from planets and satellites elsewhere in the solar system, all of which can be dated by isotopic means, we may be able to distinguish the patterns of abiological synthesis – which, although not as highly organized a process as biological synthesis, is clearly not totally random – and even detect some first drafts for life that never got off the drawing board. For example, not only have abiologically synthesized amino acids been discovered in meteorites, but also, late in 1972, two Soviet scientists, Alexander Vinogradov and Gennady Vdovykin, reported the isolation of DNA-like polymers from the Mighei meteorite, which landed in the Ukraine in the late nineteenth century.

Further knowledge in this field could also come rapidly as we explore more of our solar system, for example through experiments such as some of those carried out as part of the 1976 Viking landings on Mars, and the close-up observations of Saturn's biggest moon, Titan, planned for 1981. Titan is comparable in size to Mercury and Mars and has an atmosphere which may be as large as the Earth's. This atmosphere probably contains organic compounds, while the surface may be covered with tarry polymerization products. According to Dr Garry Hunt of the British Meteorological Office, 'The surface of Titan may be the best place in the solar system for the investigation of prebiotic organic chemistry.'[14]

14. Garry Hunt, 'The riddle of Titan's gas', *New Scientist*, lxiv, 1974, p.429

In writing about the origin and nature of life, Manfred Eigen has claimed that 'it is most characteristic of all living organisms that they are in a state far removed from chemical equilibrium'.[15] One major consequence of this is that the totality of life – the biosphere – on any planet is also likely to be out of chemical equilibrium. Working from this hypothesis in the 1960s, the British chemist James Lovelock developed a novel suggestion for detecting life on other planets: examine their atmospheres for signs of chemical disequilibrium.

As it is almost impossible to conceive of a biosphere which would not use the planet's atmosphere as a cycling system for transporting metabolic products from one life form to another – as the carbon dioxide we exhale is carried to plants for their respiration – a planet on which life exists ought to have a non-equilibrium atmosphere. So far, the only planet to fit this criterion is the Earth. However, the Lovelock technique could be used by space probes to detect planets that probably bear life. The only disadvantage is that, apart from our solar system neighbours, the astronomers have not yet been able to discover direct evidence for the existence of planets elsewhere in our galaxy . . . so the method may have to stay in cold storage for a few years.

15. Eigen, op cit.

# *CHAPTER 4*

## THE CHEMISTRY OF LIFE AND DEATH

Somehow, from the primordial slime, came man. Between the simple chemical postulates about how life might have originated and the masses of men and animals that now inhabit our planet is a great span that lies outside the realm of the chemist. Yet, if there is a dominant theme in chemistry today regarding basic research, it is the continuing and detailed analysis of the chemical make-up of man and other life forms.

If I had to pick a single event as the significant 'starter' for this chemical-biological activity, it would be the elucidation of the structure of insulin by Frederick Sanger and a few colleagues in Cambridge during the 1950s. The task took Sanger nearly ten years. He began with the then nearly brand-new technique of paper chromatography, without which his task would have been impossible. The real importance of the insulin analysis was that Sanger, by his patient and continued use of this technique, showed scientists that a whole new order of problems could be tackled. Life was still difficult to understand, but chemists could help put pieces together.

Insulin had been known since the 1920s, when George Banting and Charles Best, working in Toronto, showed that if a dog's pancreas is removed the dog develops the symptoms of a disease known to man since Greek times: diabetes mellitus. Banting and Best also showed that an extract from pancreas alleviated diabetic symptoms; the active ingredient of this extract was the hormone, insulin.[1] On the basis of this discovery was built an industry, preparing insulin from animals for the relief of humans suffering from diabetes. Yet nobody really knew what insulin was.

It was not difficult to find out that insulin is a protein, but this was not a great advance on previous knowledge. Proteins, as was explained earlier, are large molecules made up from about twenty naturally occurring amino acids. In the simplest proteins, amino acids are joined together in a single long chain. However, many proteins are made from more than one chain, the different chains being held together by sulphur-sulphur bonds

1. Hormones are compounds produced by living organisms which profoundly affect their activities, although present in only tiny amounts.

(disulphide bridges), formed from the side-chain sulphur atoms of the amino acid cysteine. To 'know' the structure of a protein, it is not sufficient to know how many of each type of amino acid is present in a single molecule; it is also necessary to know the order in which amino acids occur on any chain, how many chains there are, and at what points they are joined to one another.

In the early postwar period, it was not difficult to establish how big a protein molecule was, and how many of each type of amino acid were in it; reliable methods for molecular weight determination existed, and chromatography could identify the individual amino acids and the ratios in which they occurred. But take a small fraction of a protein molecule (called a peptide), made up of a chain of four amino acids. If simple analysis shows that the four are all different, with only one of each in the peptide, there are four possible alternatives for first position in the chain, three alternatives for second, two for third, and one for last, giving, in all, twenty-four possible peptides. Insulin contains seventeen different amino acids and a total of fifty-one amino acid residues; how can one begin to determine their sequence?

Sanger started at the end – or rather, ends. Each amino acid molecule carries in its structure an amino group and a carboxylic group, both attached to the same carbon atom; peptide linkages form when the amino group from one and the carboxylic group from another come together in an amide bond. This means that, in a peptide or protein chain, all the amino and acid groups are taken up in bonds except for a free amino group at one end of the chain and a free carboxyl group at the other. Sanger found a chemical which reacts with protein chains by tagging on a yellow-coloured dinitrophenyl group to any free amino group. After tagging, the molecule can be broken down (hydrolyzed) to free amino acids, and the tagged amino acid identified by chromatography.

Sanger identified two such tagged amino acids, which indicated that insulin was composed of two peptide chains. In one of the chains, the N-terminal (so-called from the *N*itrogen in the amino group) amino acid was glycine. By using mild hydrolysis, it was possible to break the marked chain into small peptides, rather than individual amino acids, and then identify the peptide fragments individually by further hydrolysis. For example, one fragment hydrolyzed to two amino acids, the labelled glycine and isoleucine; another consisted of three amino acids, glycine, isoleucine and valine. Had only the tripeptide glycine-isoleucine-valine been found, it might have had two structures, with the isoleucine sandwiched between the glycine and valine, or with the valine sandwiched between glycine and isoleucine. Because Sanger had also found the dipeptide glycine-isoleucine, he knew that glycine-isoleucine-valine must be the correct structure.[2]

With the aid of colleagues Hans Tuppy and E. O. P. Thompson, Sanger worked out the complete structures of the two chains by splitting them into fragments and comparing them to see where overlaps occurred. For example, from identifying two peptides from one chain as Ser.His.Leu.Val and Leu.Val.Glu.Ala, it was possible to postulate that part of the chain sequence is Ser.His.Leu.Val.Glu.Ala. By 1952 Sanger had the structures of both chains, but as separate molecules. It still remained to work out how they were joined together. One chain contained four cysteine residues, the other two, thus making several combinations possible. Discovering which was the natural combination was difficult, for disulphide bonds have a tendency to break and re-form in a different order during experiments. Sanger finally found a way to prevent this, and identified the disulphide bridges correctly.

This was a most important step, for if the disulphide bridges are joined wrongly, the molecule shows no insulin activity. The same is true of many multichain proteins, and this raised a problem that was not solved to everyone's satisfaction until the late 1960s when, again, insulin played a part.

Proteins are one of the most important classes of molecule in living organisms. Hair, cartilage and muscle are all made of protein. Perhaps more important, so are enzymes, the catalysts which plants and animals use to speed up reactions which might otherwise take hours, days, or even years, certainly, at an uncatalyzed rate, too long for life to continue.

The major component of any enzyme is a protein. As chemists studied enzymes to find out how they worked, they noted some important lessons about chemistry. First, even if you specify the order of amino acids in a protein, it will not necessarily behave properly; it has to be folded up in the correct way. Not only does each part of the molecule have its own stereochemistry, but the overall shape that comes from putting all the pieces together in the right way is also essential. This opened up an important question: How does nature do it? When scientists broke the disulphide bonds in multichain enzymes, they could not get them to rejoin in the right order.

2. For convenience, the names of amino acids are abbreviated when the structures of peptides are described. The abbreviations for glycine, isoleucine, and valine are Gly, Ileu, Val and the peptide just referrred to is shown, in the shorthand form, as Gly.Ileu.Val. The other amino acids commonly found in proteins (with their abbreviations) are: alanine (Ala), arginine (Arg), asparagine (Asn), aspartic acid (Asp), cysteine (Cys), glutamic acid (Glu), glutamine (Gln), histidine (His), leucine (Leu), lysine (Lys), methionine (Met), phenylalanine (Phe), proline (Pro), serine (Ser), threonine (Thr), tryptophan (Try) and tyrosine (Tyr).

This spoiled a nice but tentative theory which had been put forward by scientists working with single-chain enzymes. In the late 1950s they had found that, with enzymes such as ribonuclease, it was possible to break the disulphide bonds; when they were rejoined, all the parts went back in their proper places. The theory was that nature needs no elaborate mechanism for making disulphide bonds in special ways; as a long protein chain is synthesized, it curls up in a way that is inherent in the order of the amino acids. When the time is right to form disulphide bonds, all the sulphur atoms are positioned so that only the right bridges are built. The uncooperative nature of multichain enzymes spoiled this theory. To overcome this difficulty, another theory was needed.

It had been discovered, for example, that certain digestive enzymes, such as chymotrypsin, are stored in the body in an inactive form. Chymotrypsin digests protein that we eat. If it were kept in the body with no outside protein to feed on, it would start feeding on the body itself – an unsatisfactory and unpleasant prospect. So chymotrypsin is stored as chymotrypsinogen, which contains several more amino acids in each molecule than chymotrypsin itself, but is enzymically inactive. It is also a single-chain molecule; when an active enzyme is needed, two small peptides are bitten out of the chymotrypsinogen by another enzyme. This changes the three-dimensional structure, and produces a three-chain, active enzyme. When the disulphide bonds in single-chain chymotrypsinogen are broken and rejoined, they, like those in ribonuclease, all link up in the proper places.

Because there is good reason why enzymes should have precursor molecules, the chymotrypsinogen-chymotrypsin relationship was not sufficient to establish a watertight case for believing that all multichain proteins come from single-chain precursors. Interest was aroused in looking for precursors of other proteins and, because of its importance as a therapeutic material, various groups searched for an insulin precursor. Despite a number of experiments using radioactive labelling techniques which indicated that there was no 'proinsulin', two University of Chicago scientists, Donald Steiner and Philip Oyer, decided that there probably was a single-chain insulin precursor, and published details of studies of insulin synthesis in living cells which had led them to this conclusion. Their work was read by a team, working for the Lilly Research Laboratories in Indianapolis, led by Ronald Chance. The group realized that a substance they had discovered while purifying pig insulin might be the precursor.

When it was analysed – modern techniques are much faster than Sanger's original ones – the substance from Indianapolis was found to be a single-chain protein, the ends of which were identical to the two chains of insulin, with an intermediate chunk of thirty-three amino acid residues holding them together. Proinsulin had been found. When it was treated with trypsin, the enzyme which turns chymotrypsinogen into chymotrypsin, a large section was removed from the middle to give a two-chain molecule almost identical with insulin; the only difference was that trypsin had bitten too hard, leaving one chain a single amino acid residue short.

Insulin had come to the fore once again to settle a point in biological chemistry. Few scientists now doubt that multichain proteins arise from single-chain precursors. However, this is really only a small part of the puzzle which protein chemists are trying to fit together, because accepting that the overall structure of a molecule determines its activity does not explain why it acts.

At times, knowing the exact three-dimensional structure of a protein molecule can give clues to how it works – this has been the case with some of the enzymes. All exact three-dimensional structures are obtained by X-ray diffraction studies. Use of this technique in biological studies began in the 1930s in England under the direction of J. D. Bernal. Working with him was Dorothy Hodgkin, who began work on insulin crystals in 1935. Together with a team of helpers, she published a three-dimensional structure for insulin thirty-four years later, in 1969. (She had also done much other work during this time, including the structure determination of vitamin B which won her a Nobel prize in 1964.) Unfortunately, there was nothing in the structure that hinted at the mode of action.

Another way in which clues to the mode of action of a protein can be determined is to modify it chemically. Doing this on a systematic basis means synthesizing the real substance first, after which one can perform allied but slightly different syntheses. Synthesizing a protein is no easy task. To synthesize a single dipeptide (a molecule composed of only two amino acid residues) requires between three and six distinct chemical operations. Syntheses of small peptides with biological activity were undertaken extensively during the 1950s and 1960s – oxytocin and vasopressin, two octapeptides (eight amino acid residues) from the posterior pituitary gland were the first, synthesized in 1955 by the American chemist Vincent du Vigneaud, who received a Nobel prize in chemistry for his achievement. This work not only established the methods of chemical synthesis of protein material, but also made clear that the first person (or group) to complete an actual protein synthesis would have to put in a tremendous effort, much of it in performing repetitious reactions. Three groups seemed, in the early 1960s, the only real contenders for the title, and all three were trying to synthesize insulin, because of its smallness.

A German group working at Aachen under H. Zahn synthesized one of the two chains of sheep insulin in 1963; the same chain was synthesized in Pittsburgh in the same year by P. Katsoyannis and his co-workers. In the following year, a third group, from China, synthesized the same chain, but from bovine insulin, which has a slightly different structure (the ninth amino acid in the twenty-one-link chain is different in the two species). The other chain, with thirty links, was synthesized by all three groups in 1964 (cows and sheep have identical structures for these chains). After that, it was a matter of joining the disulphide bonds together in the right way, and in November 1965 the Chinese group, composed of more than a score of scientists from the Academia Sinica in Shanghai and Peking University, published results of their success. Their yield – the quantity of material actually made, as compared with the amount which can theoretically be derived from the quantities of starting material used – was only about 2 per cent. None the less, man had made a protein in the laboratory.

This feat has since been repeated with larger proteins, perhaps most spectacularly with the synthesis of an enzyme, ribonuclease. At the beginning of 1969, two American groups announced simultaneously that they had achieved synthesis of this single chain of more than 100 amino acid residues. One group, working at the American pharmaceutical company Merck, Sharp & Dohme and led by R. Denkewalter and R. Hirschmann, used a traditional method; the other group, headed by Bruce Merrifield of Rockefeller University, used an automated approach.

Merrifield developed his method in 1959. It does not reduce the number of chemical operations in a peptide synthesis, but it programmes them so they can be automated. Peptides are synthesized in solution or liquid suspension. Merrifield's basic idea was to attach the first amino acid in his syntheses to a polymer which is available in the form of small beads. He then arranged the various chemical procedures of peptide synthesis in such a way that, at the end of each peptide bond-forming step, everything but the polymer beads was thrown away. The addition and removal of solutions was readily automated to produce the 'solid phase peptide synthesis' method, with necklaces of peptide chains growing on insoluble supports.

Since the success of this method in enzyme synthesis, it has been taken up widely and improved upon by protein chemists. One disadvantage of the method is that, at any time, thousands of peptide chains are growing on the polymer support. When an amino acid is introduced to add on to these chains, not all of them may react (i.e., the yield may be less than 100 per cent). When the sequence is repeated, the same thing will happen again, but it is not likely that those which did not react at the earlier stage will be unreactive again. Therefore, among the correct length protein chains – in which reaction has occurred at each step – will be a number of 'sports', short chains which have stopped reacting for some reason, or chains in which one or more residues have been missed out. In 1969 a group led by Professor T. Wieland at the Max Planck Institute for Medical Research, Heidelberg, developed an additional step in the process which ensures that all unreacted chains, at any stage of the synthesis, are stopped off by addition of a molecular fragment that prevents further reaction and makes these necklaces easy to separate from the required material.

Despite Merrifield's automatic procedure, the manufacture of proteins is not easy. So why do it, when supplies of needed proteins – such as insulin – can be obtained from animal sources? One reason is because it enables man to make unnatural proteins and to study the effects of changes in the composition of the biological activity of a molecule, for the delicate relationship between a molecule's shape and its activity is not shown only by the need to have disulphide bonds joined correctly.

## What will proteins tolerate?

In 1947 a form of anaemia which still kills about 80,000 children annually was shown to be due to a fault in the haemoglobin of those who suffered from the disease. The fault was easily recognized; under the microscope their venous blood could be seen to contain deformed red cells. Because of their sickle-like shape, the disease is called sickle-cell anaemia. Haemoglobin is a protein-iron complex responsible for taking oxygen from the lungs and distributing it around the body via the blood stream; sickle-cell anaemia kills because the victim is incapable of supplying sufficient oxygen to his tissues.

In 1953, Vernon Ingram, working in the Laboratory of Molecular Biology at Cambridge, discovered the reason for this. He found that the difference between sickle-cell haemoglobin and ordinary haemoglobin is no greater than two amino acid residues. Since haemoglobin is made up from four protein chains, in two identical pairs, the change which kills is a single amino acid substitution (glutamic acid replaces valine) in a protein made up from more than 140 amino acid residues.

Before detailing the story of haemoglobin further – and it is mainly the story of thirty years in the life of one man, Max Perutz – let us turn to the more general question of protein function. The story of the sickle-cell can be misleading. Certain parts of a protein molecule are crucial, while some protein molecules can lose large chunks without losing their activity.

What became clear during the 1950s and 1960s was that, in many enzymes, the chemical substances whose reactions are catalyzed (known as substrates) come into contact with very few of the thousands of atoms in the enzyme. In 1960, one of the most prolific theoreticians of enzyme chemistry and biology, Daniel Koshland, Jr., of the University of California, Berkeley, proposed a fourfold classification for the amino acid residues in a protein: 'contact' residues, those that come into contact with the substrate, with the possibility of orbital overlap and bond formation; 'auxiliary' residues, which contribute to the flexibility of the 'active site', as the sum of the contact residues is called; 'contributing' amino acids, which maintain the structure of the enzyme in the active configuration; and 'non-contributing' residues, which can be removed without loss of activity. In most cases, the contact residues are only two or three out of the hundred or more residues in an enzyme. How did the concept of such a small active site, clearly of great importance in elucidating the chemistry of life, arise? It came from the chemistry of death – in the shape of nerve gases.

Shortly before World War II, a group of organophosphorus compounds were discovered which had potent effects on living organisms; during the war, compounds in this group were manufactured as nerve gases. Fortunately, mutual fear kept both sides from unleashing chemical and biological weapons, so the gases were never used. But scientists and governments throughout the world were anxious to know how these agents worked, partly because, without an understanding of the mechanism of action, antidotes could not be designed.

It was discovered that the organophosphorus compounds attack an enzyme, acetylcholinesterase, which plays a crucial role in the functioning of the nervous system. The enzyme and the nerve gas react to form a new compound which does not have the enzyme's essential biological activity.

Acetylcholinesterase is not the only enzyme inhibited by the organophosphorus compounds; another is that well-studied digestive enzyme, chymotrypsin. When organophosphorus-treated acetylcholinesterase, chymotrypsin, and various other 'inhibited' enzymes were analysed, the phosphorus atom of the nerve gas was always found chemically bonded to the side-chain of a single residue of the amino acid serine. This reaction was quite specific. Despite the presence of up to a score of serine residues in the enzyme, only one – and always the same residue – was phosphorylated. From this result the idea of the active site developed; for some reason, some residues in an enzyme are more active than others. In the last fifteen to twenty years, scientists in many countries have designed a variety of active site inhibitors, small molecules which will inhibit enzymes and leave part of themselves attached to a single residue, labelling it as part of the active site of that enzyme.

The reason why two or three amino acid residues, not necessarily contiguous on a protein chain, form an active site becomes clear from three-dimensional studies of enzymes. These show that the active site is an actual physical site with a shape. The residues that compose it are brought close together by bends, folds and loops in the protein, forming a cleft or cavity into which substrate molecules will just fit. Some families of enzymes have the same residues making up their active sites, but slightly different conformations alter the specificity of the enzyme, so that although the same reaction is carried out by the whole family, individual enzymes catalyse it in different molecules.

An example is provided by the four major protein-digesting enzymes (proteases) in humans: elastase, trypsin and two sorts of chymotrypsin. All of these catalyse the rupture of peptide bonds, but each prefers to break peptide bonds in which particular amino acid residues are involved. Recent research on their three-dimensional structures, together with sequence determinations, has led to the view that there was originally only one enzyme. Like all major proteins, these enzymes are synthesized by a method dependent upon hereditary information, the genetic code (see p. 110).

An evolutionary process for which there is strong evidence is 'gene doubling', in which a gene suddenly doubles so that two genes are carrying the same code (for an enzyme protein sequence, for example). As the organism in which this gene doubling occurs evolves, the genes mutate, so that the enzymes produced are no longer identical. Close similarities among the four digestive enzymes are indicative of gene doubling followed by divergent evolution. Obviously, if a gene diverges too much, it may eventually code for a useless molecule without biological activity. Alternatively, a totally new biological function may appear. Before exploring this angle, convergent evolution, which does the same thing in different ways, deserves mention.

Some bacterial proteases developed separately from mammalian proteases and have totally different structures, yet perform the same function – the breaking of peptide bonds. A bacterial protease sequence worked out in 1970 in Canada showed sufficient similarities of structure to mammalian proteases to postulate evolution from a single ancestral gene, but this is not the case with better-known bacterial proteases such as subtilisin. Although both mammalian and bacterial proteases have a serine residue in the active site, no comparisons of sequence of the type needed to postulate that there was once a common gene can be made. This, as Ottawa chemist Bernard Belleau showed in 1970, does not mean that the enzymes necessarily have differently shaped active sites.

Belleau synthesized an artificial substrate molecule which could be broken down by a wide range of proteases. The molecule was flexible, able to accommodate its shape to differences in shape of the different active sites. Belleau then modified the molecule so that it was no longer flexible, and had a precise shape. This did not react when placed in contact with most of the enzymes used. Only two of them broke it down: chymotrypsin and subtilisin. In both cases the substrate was hydrolyzed at the same rate. The conclusion is that subtilisin and chymotrypsin have identical three-dimensional spaces in their active site regions, despite the use of different building blocks in the composition of these spaces.

Now to return to divergent evolution. As more and more biologically active peptide sequences have been determined and computer comparison techniques developed, it has become clear that gene doubling, followed by complete divergence of function, has occurred on numerous occasions during evolution. As with the digestive enzymes, divergence of function is not always complete. Among the hormones involved in human alimentary operations are secretin and glucagon, responsible, respectively, for causing the pancreas to produce digestive juices, and for breaking down glycogen – a glucose polymer used by the body as an energy store – into individual glucose molecules. Despite their different effects, the two hormones are both about the same size (twenty-seven and twenty-nine amino acid chains respectively) and half of the peptide sequence is the same in both molecules. Two other peptide hormones which also share parts of their structure are melanocyte-stimulating hormone (MSH) and corticotrophin, yet one is responsible for controlling skin pigmentation, while the other regulates the activity of the adrenal cortex, itself responsible for producing the steroid hormones, which play a variety of roles in the body.

When did these diversions happen? Is it possible to use the knowledge of chemical structures now available to improve our understanding of history? The answer is a qualified yes. A number of proteins are common to many species of animals, and some are found in animals, plants and micro-organisms. As more of these proteins are purified and sequence-analysed, it is possible, with the aid of modern computing techniques, to compare the same protein from different species and to chart its evolution. For example, when the sequence of haemoglobin was compared among a few species, it was found that there was no difference between haemoglobin proteins in man and chimpanzees. Gorilla haemoglobin differed from human by two amino acid residues, while that of man and monkeys differed by twelve. If it is assumed that, once a species splits into two, the rate at which mutations change the structures of key molecules is roughly constant, we can draw up a scale of 'closeness' between species in evolutionary history, showing how far back ancestry was common.

The respiratory protein cytochrome c has been studied extensively in

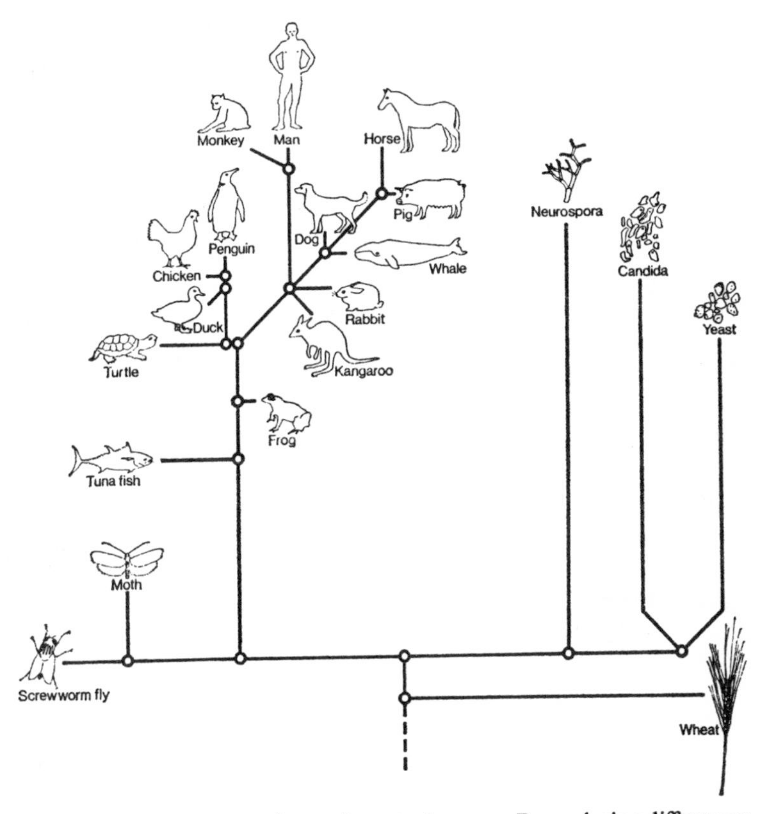


FIGURE 11 A phylogenetic tree for cytochrome c. By analysing differences in the structures of cytochrome c – a respiratory protein – from different species, it was possible to generate this evolutionary tree, using a computer programme.

different species. It is a single chain of about 100 amino acids, and several dozen species of the protein have been sequenced. Figure 11 shows a phylogenetic tree drawn from data on some of these. In order to insert an absolute time scale it is necessary to use evidence from other sources which pinpoints the divergences of groups from each other. But, accepting the linear rate of mutation, a fixed external point of reference makes it possible to use this method to go back beyond the earliest external reference.

However, we cannot assume that different proteins will have the same rate of change. The evidence at hand consists solely of functional proteins. Protein mutation may take place at a constant rate, but it will produce many more non-functional proteins than functional ones. Lethal mutations leave no evidence, so, indirectly, interspecies comparison of proteins also tells something about the structure-function relationship. In cytochrome c from all the species studied, there are thirty-one invariant amino acids; presumably the animals, plants, or microbes in which any of these thirty-one amino acids mutated found themselves with non-functional cytochrome c – or, rather, did not find themselves anywhere, as they were dead.

A group of proteins called histones have been partially sequenced, and it appears that one of them, histone IV, from cattle and peas, has undergone only two substitutions since these two groups of organism diverged $10^9$ years ago. Obviously the histone IV molecule is built to rigorous specifications and will not work if many changes are made.

Another ambiguity in this type of evolutionary dating is the possibility of back mutations. One amino acid may be substituted by another but, given a further mutation, the original sequence may recur. Similarly, once two species have diverged, a protein common to them may undergo identical mutations, so that both make the same amino acid substitution. In compiling phylogenetic trees such as that for cytochrome c, the computer programme takes such possibilities into account.

On the surface, the likelihood of a back mutation seems small. There are twenty different amino acids composing protein chains, which gives a low probability of any particular substitution. However, this is only a superficial appearance, for it is not the proteins that mutate. Although proteins play a vital role in living organisms, they are not the master molecules of life. This title is taken by the nucleic acids, which carry the commands for all the living activities of a cell, and pass those commands on between generations.

## Master molecules

The story of the unravelling of the genetic code is complex and exciting, for the science of molecular biology which arose from it has exercised some of the most fertile minds of the postwar generation in a variety of scientific disciplines. As this book is primarily concerned with chemistry, it can only look at part of this story – the actual molecular structures and what they do. How they do it is beyond our scope.

There is a basic difference between nucleic acids and the other macromolecules mentioned in Chapter 3. Proteins are made by linking together amino acids, carbohydrates by linking together sugar molecules. The basic unit for making a nucleic acid is not, like these, a simple molecule, but a more complex entity called a nucleotide which is derived from three simpler molecules. According to whether the sugar part of the nucleotide is ribose or deoxyribose, the resultant nucleic acids are called ribo- or deoxyribonucleic acids, abbreviated to RNA and DNA. The fundamental parts of both DNA and RNA are the nitrogen-containing organic bases attached to the sugar residues. These are of two types, the purines adenine and guanine, and the pyrimidines cytosine, thymine and uracil, which is found in RNA in place of thymine. (A number of other bases, slight chemical modifications of these five, occur occasionally in RNA.) In referring to nucleic acid sequences, the first letter of each base is used as an abbreviation.

It was in 1953 that James Watson and Francis Crick proposed the structure of the hereditary DNA which is now known as the double helix. In a single, elegant concept, Crick and Watson showed how heredity was possible at the molecular level.

In part their hypothesis depended on the observation by the Austrian-American biochemist Erwin Chargaff that the adenine:thymine and guanine:cytosine ratios in DNA samples were always close to unity. What Crick and Watson suggested was that, in a gene, two long, fibrous molecules of DNA twine around each other in a complementary fashion. Purine bases (A and G) are larger than pyrimidine bases (C and T); when two DNA strands form a helix, the bases point inward and a large base on one chain forms a snug fit with a small base on the complementary chain. This 'base pairing' of A with T and G with C involves substantial hydrogen bonding, which stabilizes the double helix. When the two complementary strands of a double helix are separated, it is possible for new DNA synthesis to reproduce the missing half of each helix. In this way the DNA forms a 'code' which can be copied. This is the basis of heredity: DNA stores the information that makes us man or mouse, and ensures that our children are the same.

Through a complex and incompletely understood process, DNA strands are used in cells to synthesize RNA, and it is from the code inscribed on the RNA that cellular particles called ribosomes manufacture proteins. How does this transcription occur?

As it is an information source, the genetic code must resemble information communication systems with which we are familiar, such as words, sentences and books. The obvious analogy is to suggest that the four bases of DNA are the letters of the alphabet, and the order of these makes up 'words'. Although this simple idea is also correct, it does not mean that the code is solved. Achievement of this task in the late 1960s was the result of years of complex experimentation by many researchers.

The 'language' of the genetic code need have only twenty words, for this is sufficient information to specify all the amino acids that make up proteins. Obviously, neither a single base nor a pair of bases are sufficient as the codewords for proteins, for these provide only four or sixteen possibilities, respectively. But a code which uses a sequence of three bases to specify an amino acid permits sixty-four possibilities. The correctness of the three-base (triplet) code hypothesis was substantially shown by the work of Crick and his collaborators on the DNA of a bacteriophage – a virus which attacks a bacterium – known as T4. It was found possible to mutate this virus selectively, to treat its DNA in such a way that a number of different, defined effects might take place.

In a sentence written in the English language, we show where a word ends and another begins by leaving a space between them. This is a convention rather than a necessity; the Romans, for example, did not usually leave spaces between words. However, where words differ in length it is clearly advantageous to signal beginnings and endings. The words of the genetic code are all the same length – three letters – and are made up into sentences with no spaces between them; obviously the mechanism which 'reads' the code knows the length of the words and behaves accordingly. That this is the case was shown by Crick's mutations which in some cases deleted or added single bases to the bacteriophage DNA; in other cases, two or three bases were deleted or added. Unlike English, nearly all the possible permutations in the genetic language make sense. Although there are only twenty amino acids, most possible triplets 'code' for one of them (in other words, some amino acids correspond to more than one triplet; the genetic code contains synonyms). If, for example, we delete one letter (in this case the third) from the triplet sentence, *The man was fat and bad*, and then read off the triplets, we get nonsense: *Thm anw asf ata ndb ad.* With the synonymous genetic code, the chances are that a protein will still be formed on decoding through the RNA mechanism, but from the deletion onward it will have an amino acid sequence different from the protein produced from unmutated DNA. However, if a triplet of bases is removed or inserted, the protein produced will differ from normal protein only in having one amino acid less or more. Crick's experiments indicated that this did in fact happen, and clinched the triplet as the basic word length in the genetic code.

Once this had been done, everyone wanted to know what each word spelled out. Possibly the major contribution of pure chemistry to molecular biology was the work that made it possible to find out. This work could not have been accomplished without the help of biologists who developed cell-free preparations of living material capable of carrying out protein synthesis in test tubes.

In the early 1960s Marshall Nirenberg and his collaborators at the National Heart Institute, Bethesda, Maryland, decided to study the behaviour of cell-free systems when presented with some synthetic RNA. This synthetic RNA, made with the aid of an enzyme discovered by Severo Ochoa of New York University, was of several types; the sequence in the simplest contained only a single base, such as uracil. Using an all-U-polymer, Nirenberg found that a cell-free system produced a polymer of just one amino acid: phenylalanine. The first of the sixty-four possible triplets, UUU, had been decoded.

In addition to using other single-base RNAs, Nirenberg's group also used RNA made from known mixtures of bases. If, for example, RNA is made using equal parts of U and A as the only two bases, statistical calculations can decipher how frequently each of the possible triplets (such as AAA, AUA, UUA) will occur. By checking these probabilities against the actual incorporation of amino acids, it was possible to assign a number of other triplets to specific amino acids.

While this work was in progress, Professor H. G. Khorana at the Institute of Enzyme Research, Madison, Wisconsin, an Indian-born chemist who had studied at Cambridge under Nobel prizewinner Lord Todd, was perfecting methods for the chemical synthesis of short strands of DNA with completely specific structures. Using the appropriate starting materials, Khorana made the dinucleotide AC; he then polymerized this to a twelve-base chain ACACACACACAC. By the same technique he synthesized the complementary sequence TGTGTGTGTGTG, and the two combined to make a small strand of double helix. With the aid of an enzyme called DNA polymerase, Khorana found that this small sequence could be polymerized into a double helix containing about 3,000 bases in which the order of the initial dinucleotide was maintained.

Khorana extended the method by making his initial synthesis more complex and producing tri- and tetranucleotides of known sequence. In all he produced eight 'artificial' DNAs. When these were presented to cell-free synthesizing systems, they produced RNA, which produced protein, which unambiguously confirmed the triplet nature of the code, and its translation. For example, take a dinucleotide based polymer, AC/TG. When this is converted to RNA by transcription, it produces a repeating sequence of either UG or AC. Either of these polymers, UGUGUGU. . . and ACACACAC. . . , can be read in two ways as triplets (UGU/GUG, ACA/CAC), and both were found to produce proteins made up of two alternating amino acids: threonine and histidine for ACACAC . . . and valine and cysteine from UGUGUG. . . .

There are two other ways in which the code resembles language with which we are familiar. When we end a sentence, we mark the fact with a full stop. Two of the triplets of bases, rather than coding for amino acids, are full stops. When the protein-synthesizing machinery reaches this part of an RNA molecule, it ceases to manufacture the protein and instead releases it.

The other important task that must be faced is beginning a sentence – or a protein. In the genetic code this is most important for, unless the synthetic machinery knows exactly where to begin, it may start reading from the wrong base and produce a completely nonsensical (non-functioning) protein. At the moment it seems that the codeword for the amino acid methionine (AUG) also acts as a chain-starter. Where this is so, formylmethionine is coded for; its amino group is blocked, so that the protein can only be made in one direction. If amino acids are joined together in sequence, even if the addition of amino acids is given in a fixed order there are still two possible sequences, according to whether the second and subsequent amino acids link on to the chain through their amino or their acid groups. By blocking the amino group of methionine with a formyl group, this ambiguity is removed.

It is believed that all protein syntheses start with methionine, although in many cases this residue, and perhaps a bit more, is lopped off at some stage before the protein reaches its final form. The genetic code sequences are shown in Table 3.

From this it can be seen exactly how a 'point' mutation – that of a single base – can be deleterious. To take a realistic example, consider a short segment of double helix with the sequence

...TGA CTC AGG TAA...
...ACT GAG TCC ATT...

The simple molecule nitrous acid ($HNO_2$) is a powerful mutagen because it reacts with amino groups, and adenine, guanine and cytosine all have free amino groups. They are not all equally reactive toward nitrous acid, and reactivity of individual residues in a DNA molecule differs from reactivity of the isolated base, for the shape of the macromolecule protects some amino groups while tending to expose others. Adenine and guanine are deaminated by $HNO_2$ to two other purine bases, hypoxanthine and xanthine (X), while cytosine is deaminated to the base common in RNA, uracil. Let us assume that, in the sequence above, $HNO_2$ attacks all the guanine and cytosine residues. Then, the sequence is changed chemically to:

...TXA UTU AXX TAA...
...AUT XAX TUU ATT...

TABLE 3 *The genetic code*

| First letter | Second letter U | C | A | G | Third letter |
|---|---|---|---|---|---|
| | phe | ser | tyr | cys | U |
| U | phe | ser | tyr | cys | C |
| | leu | ser | END | NONSENSE | A |
| | leu | ser | END | try | G |
| | leu | pro | his | arg | U |
| | leu | pro | his | arg | C |
| C | leu | pro | gln | arg | A |
| | leu | pro | gln | arg | G |
| | ileu | thr | asn | ser | U |
| | ileu | thr | asn | ser | C |
| A | ileu | thr | lys | arg | A |
| | met (BEGIN) | thr | lys | arg | G |
| | val | ala | asp | gly | U |
| G | val | ala | asp | gly | C |
| | val | ala | glu | gly | A |
| | val | ala | glu | gly | G |

If, in both cases, we assume it is the top strand which acts as a template for RNA, the transcription sequences will be:

...TGA CTC AGG TAA... ...TXA UTU AXX TAA...
...ACU GAG UCC AUU... ...ACU AAA UCC AUU...

In the second case, the wrong RNA is produced because the mutated bases X and U are not complementary to C and G, as was the case originally. When the new RNA is translated into protein, it produces the sequence Thr.Lys.Ser.Ile, rather than the required Thr.Glu.Ser.Ile. Only one amino acid has changed, admittedly, but the change could have far-reaching effects, for while the side-chain of glutamic acid is acidic, that of lysine is not, and such a change could be sufficient to alter the overall shape of the whole protein and make it non-functional.

Earlier in this chapter it was remarked that the disease sickle-cell anaemia is caused by a single amino acid substitution: glutamic acid instead of valine. The code for valine is GU(*X*) (where (*X*) can be any of the four RNA bases); glutamic acid is GAA or GAG. Consequently, this lethal change can be attributed to a single change in DNA that results in the messenger RNA containing the base A at one position rather than U.

This is also why some of the complexities of molecular evolution are as great as they are, for a back mutation may require only a single base change. Instead of selecting from twenty possible amino acids, for mutation's sake selection is made from only four bases. Another reason for expressing caution about molecular genealogy is that the degeneracy of the code can lead to hidden mutations. If we find a sequence of RNA coding for an amino acid – for example, UCU (serine) – we have no way of knowing whether, at some time past, that part of the RNA did not read UCC, which also codes for serine. Nor, at present, can we tell how many of the codons of different RNAs coding for the 'identical' portions of cytochrome c molecules are themselves identical. One reason for the degeneracy of the genetic code may be that it confers survival value: it can mutate within wide limits without producing dysfunctional molecules.

### Breaking or making a gene

In addition to the messenger RNA, which carries the coded message from DNA to the ribosome, which reads it and converts it into protein, there is another form of RNA involved in protein synthesis. These are the relatively small molecules, known as transfer-RNAs (t-RNAs), which were mentioned in the previous chapter. They are responsible for lining up amino acids so that they are in the correct place on the assembly line when the ribosomal machinery reaches them. There are probably more than twenty different t-RNAs, although these are all that are theoretically needed for each of the twenty different amino acids to have its own.

In 1965 Robert Holley and his collaborators at Cornell University worked out the sequences of nucleotides in the alanine t-RNA from yeast. This research – leading to a known sequence of seventy-seven nucleotides – took seven years, much of the time being occupied in obtaining sufficiently pure material to work with, using the technique of countercurrent distribution invented in the early 1940s by Lyman C. Craig.

Holley's method of analysis involved using an enzyme to break down the t-RNA molecule into fragments which could be separated from one another by passing them, in solution, down columns of chemically modified cellulose. In many ways the sequencing method resembled that used by Sanger on insulin a decade earlier, except that the spectroscopic and chromatographic techniques available were substantially better.

Holley's t-RNA contains several unusual bases, such as methylated guanines and dihydrouridine. The most interesting aspect of the structure, however, is its internal complementarity. Parts of the chain are able to form regions of double helix with other parts. Since Holley's success, several other t-RNAs have been sequenced, and in August 1974, groups of scientists in Cambridge, England, and Cambridge, Massachusetts, published detailed analyses of the full three-dimensional structures of a t-RNA. However, during the early 1970s, the race was on for a bigger prize – total sequencing of a genome.[3]

At about the same time that Holley completed his sequence analysis of alanyl t-RNA, Fred Sanger took up RNA-sequencing studies. In 1968 he published the structure of a 120-sequence RNA known as 5S RNA, which can be isolated from ribosomes, but whose function is not understood. Sanger used a combination of electrophoresis and radiotracers.

Electrophoresis uses slight differences in the rates of movement of different molecules in solution in an electric field to help separate and identify them. It is, in many respects, similar to paper chromatography, and an outgrowth from it. By using radioactive phosphorus in such a way that it is incorporated into the molecular fragments he is studying, Sanger is able to develop his electrophoretograms photographically, and from this identify substances by their position. Once simple nucleotides are identified under the experimental conditions, it is possible to use them as the basis for a grid which automatically identifies the other spots on the photographic plate. In this way small sequences are established and, using the overlap comparison method, built into bigger ones.

Following his first foray into the field, Sanger, with the help of a number of co-workers, decided to tackle the RNA from a virus called R17. This consists of a sequence of 3,300 nucleotides, and is the virus's genetic material (some viruses use RNA in place of DNA as their information storehouse). Viruses are very simple creatures. Perhaps they are not creatures at all, for they live only in the presence of life by taking over the synthetic machinery of other cells and making it translate their information rather than its own. RNA from R17 codes for three proteins. These are the coat protein, which in the complete virus forms a protective coat around its RNA, a replicase enzyme which participates in the reproduction of the RNA, and another protein whose function is not yet fully understood.

When Sanger and his colleagues began work on this viral RNA, only the structure of the coat protein was known. Fortunately, the initial twenty-one-base RNA fragment which they sequenced coded for a short segment of this. By May 1969, when he lectured to the Royal Society in

3. A gene is that part of a DNA or – in some organisms – RNA molecule which codes for a single protein. The total genetic content of an organism is called its genome. It is now believed that segments of the genome may not code for proteins, but have some other, as yet not understood function. It has been estimated that the common intestinal bacterium, Escherichia coli, probably has about 2500 genes. The simplest known virus has three.

London, Sanger was able to give the structure of a fifty-seven-base sequence that coded for nearly twenty of the coat protein amino acids; it was the first direct confirmation that the genetic code really does work in nature in the way that had been postulated from laboratory studies. (There was no real doubt about this, but it was pleasant to have the confirmation.)

Rather like Holley's t-RNA, this fragment had a structure capable of forming a helix with itself. A year after his lecture, Sanger had two other sequences, of forty-four and fifty-four bases, which showed a similar property. He and his colleagues had also confirmed another small point – the genetic code is itself degenerate. In their sequences they had found all six possible triplets for serine. From this work came another postulate as to why the code might be degenerate. The two other large fragments also showed possibilities of self-helification, and if self-helification is important in the viral RNA it may be necessary for degeneracy to occur in some parts in order to form a comfortable double helix fit with other parts.

Sanger returned to the Royal Society in May 1975, to deliver the Croonian lecture. On that occasion he described sequencing work which his Cambridge group had been doing on the DNA from a bacteriophage called $\phi$X174. This single-stranded, circular DNA contains about 5,375 nucleotides, and codes for 9 proteins. Sanger's group have developed even more sophisticated methods of analysis for dealing with this material, some of which provide electrophoretograms from which it is possible to read off a sequence of 60–70 nucleotide bases only a few days after analysis has begun. One interesting result which Sanger announced during this lecture was that, in the nucleotides sequenced at that time (just under 10 per cent of the total), there was a preponderance of codons with U in the third position – in marked contrast to the degeneracy of the viral RNAs. Sanger believed that such a result was not just chance, but said it was not yet possible to suggest a reason for it.

Early in 1977, Sanger's group published the full sequence of $\phi$X174. One of the most interesting aspects of its structure is that, in some parts, more than one gene occupies the same stretch of DNA; however, when these genes are 'expressed', that is, used to produce protein, the triplets are read in different phases. This dual purpose nature of part of the DNA may, of course, account for the lack of degeneracy mentioned above.

At the other pole from analysis is synthesis. While work goes on to sequence nucleic acids, there is an equally determined research effort to synthesize them. Here Khorana's name crops up again. Synthesis of a gene is a logical extension of his earlier work on nucleotide synthesis which helped unravel the language of the triplets. This does not mean that the synthesis of long nucleotides of predetermined sequence is easy.

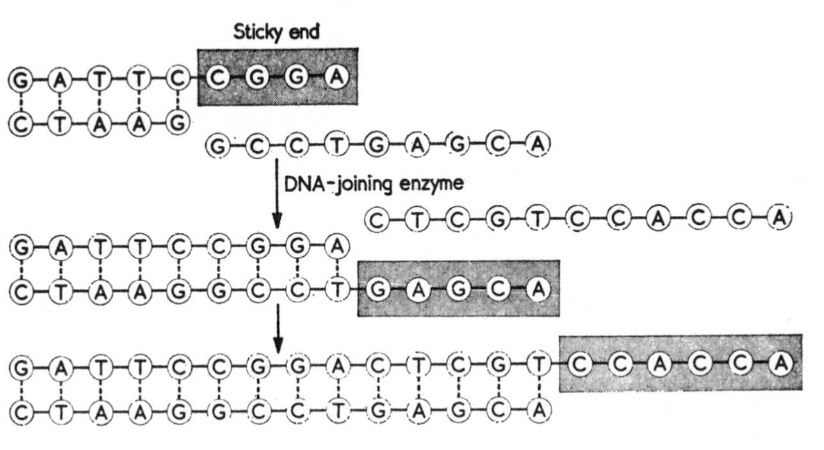


FIGURE 12 Khorana's 'sticky-end' technique for polynucleotide synthesis. Using DNA-joining enzyme, short segments of nucleic acid can be linked together, provided that a short piece of the complementary strand is available to act as an anvil for linking the chain.

In the summer of 1968, when he addressed the Fifth International Symposium on the Chemistry of Natural Products in London, Khorana said, with obvious relief, 'We have synthesized an eicosanucleotide and we shall never have to make one that long again.' He was able to say that this twenty-base chain was the largest that needed to be synthesized by purely chemical methods because of the discovery a year before of DNA-joining enzyme. Just as Khorana had recruited an enzyme to help his earlier work on producing the controlled repeating order polymers, he commandeered DNA-joining enzyme for use in what he calls 'the sticky-end technique' of DNA synthesis.

DNA-joining enzyme will link two nucleotide chains together, as long as a complementary strand overlapping the joining point on both sides is present. It is as if the second strand acted as an anvil on which the two ends can be joined together, with the enzyme acting as a hammer. As the anvil need be only four or five base units in length, use of the sticky-end technique means that pure chemosynthesis need only extend to nucleotides about ten units in length, as Figure 12 indicates.

This illustration shows double-stranded DNA in which the bottom strand is part of the code for the synthesis of alanyl t-RNA, the molecule sequenced by Holley in 1965. By June 1970, Khorana and his colleagues had completed manufacture of this DNA. A gene had been synthesized.

Just more than six years later, Khorana – now leading a large team of researchers at the Massachusetts Institute of Technology – published the synthesis of a second gene, including not only the part actually transcribed to make a t-RNA, but also the 'control' sequences at each end, which govern transcription. The synthetic gene showed full biological activity when inserted into a living cell.

Unravelling the methods which ensure that a gene is transcribed as and when required, and only then, is one of the most exciting challenges of molecular biology today. The genetic complement of most living organisms is large: the genome contains thousands of different genes. In principle, for example, each cell in a human body is capable of growing into another identical human being. All the necessary code-words are there, but most of the transcribing mechanism is switched off. In recent years there has been considerable progress toward discovering how genes are switched on and off.

It is now known, for example, that the steroid hormones which, among other things, control the development of secondary sexual characteristics such as beard growth, interact with specific protein molecules in certain cells. The complex formed between steroid molecule and 'receptor' protein in the 'target' cell then acts directly on the cell's genetic material to switch on part of the programme.

The ways in which gene expression is controlled may be of fundamental importance in the future of biology. For example, on the basis of divergences in the sequences of some of their proteins, it has been argued that 'the genetic distance between humans and the chimpanzees is probably too small to account for their substantial organismal differences'. Early in 1975, the authors of that statement suggested that 'evolutionary changes in anatomy and way of life are more often based on changes in the mechanisms controlling the expression of genes than on sequence changes in proteins'.[4]

An interesting sidelight on this issue is the case of the North American tree frogs. This concerned a new population of tree frogs discovered in Arizona. On the basis of looks, one taxonomist declared these frogs to be closely related to Mexican tree frogs; another taxonomist said that Californian tree frogs were the Arizonans' closest relatives. On the basis of similarities between proteins, the Mexican claim was upheld, and it was argued that the Arizonan frogs looked more like the Californian ones as a result of convergent evolution. This seems to show a possible opposite side of the coin to the man/chimpanzee relationship, suggesting that regulation

4. Mary-Claire King and A. C. Wilson, 'Evolution at two levels in humans and chimpanzees', *Science*, clxxxviii, 1975, p.107.

of gene expression may help different species look more alike than they really are, as well as less alike.

**Haemoglobin: iron in the soul**

So far, looking at life on the molecular level, we have identified the following pattern. At the heart lie the nucleic acids, which carry the coded sequences for and participate in the synthesis of proteins. The proteins, as well as providing structural materials, such as skin and horn, are responsible for the making and breaking of all the other molecules of life – the polymeric carbohydrates, the smaller lipids, and many other classes of compounds. This story is correct in outline, but it is not the whole story. The protein coding inherent in DNA only specifies amino acid sequences, and many enzymes and related substances are more than just proteins. In addition to the organic chemistry of life, there is an important area of inorganic chemistry (increasingly called bioinorganic chemistry), which perhaps cannot be better exemplified than by returning for a more detailed look at haemoglobin.

Haemoglobin is the material in the red blood cells of higher mammals which transports oxygen around the body from its pickup point in the lungs. Indirectly, it also removes the carbon dioxide formed by cellular processes to the lungs, where it is expelled. As the earlier discussion of sickle-cell anaemia made clear, a large part of the haemoglobin molecule is protein.

Unravelling of the molecule's three-dimensional structure was primarily the work of one man, Max Perutz. In an article published in 1964, Perutz wrote: 'In 1937, a year after I entered the University of Cambridge as a graduate student (he had left his native Austria as a refugee from Nazism), I chose the X-ray analysis of haemoglobin, the oxygen-bearing protein of the blood, as the subject of my research. Fortunately the examiners of my doctoral thesis did not insist on a determination of the structure, otherwise I should have had to remain a graduate student for twenty-three years.'[5]

The major breakthrough in Perutz's work was his discovery in 1953 of the technique of 'isomorphous replacement'. This involves adding heavy metal atoms to a crystal without altering its structure (isomorphous means having the same form). It is then possible to relate the diffraction patterns obtained by X-ray bombardment of a complex crystalline compound with individual atoms in the molecule. The first concrete result was a low resolution map of the simpler oxygen-transport compound myoglobin, obtained

5. M. F. Perutz, 'The haemoglobin molecule', *Scientific American*, ccxi, 5, 1964, p.64.

in 1957 by John Kendrew, also working at Cambridge. (Perutz and Kendrew shared the 1962 Nobel prize in chemistry for their work. In the same year Crick, Watson and Maurice Wilkins of King's College, London – who had studied the structure of DNA by X-ray crystallography – received the Nobel prize for physiology and medicine. It was a good year for molecular biologists.)

Haemoglobin is more complex than myoglobin, which is made up of only one protein chain. It consists of four protein units, composed of two identical pairs, called alpha and beta. The four 'globin' chains form a roughly spherical molecule, with each individual globin acting as a basket to hold the 'haem' portion of the molecule, which is an iron atom and a complex, but quite small, organic structure called a porphyrin. The iron is chemically bonded to the porphyrin and to a nitrogen in a histidine residue of a globin chain (see Figure 13).

The bonding of iron in venous haemoglobin is called five-coordinate, because the iron atom is joined by chemical bonds to five other atoms. Inorganic chemists find the five-coordinate form of iron odd, for straightforward inorganic compounds of iron do not usually adopt such a structure. In recent years, however, it has become clear that five-coordinate structures of this kind are involved in inorganic chemical reactions as transitional intermediates. In other words, they are the structures which occupy the top of the activation energy barrier. This may account in part for the activity of haemoglobin: it is permanently in an activated state, and has negligible activation energy to overcome before forming an additional bond with an oxygen molecule.

This is only part of the story. Evolution has led to a molecule with more than a single cunning design feature. Why, for example, should haemoglobin have evolved as a multifunctional molecule with four oxygen binding sites? The answer is that it can function more effectively. If a molecule of oxygen approaches two molecules of haemoglobin, one of which has already taken up three molecules of oxygen, the other none, the chances are seventy to one in favour of the triply-bound molecule grabbing yet another oxygen, while the other is still left with none. This loading mechanism also works in reverse. Once a four-oxygen carrying haemoglobin loses one oxygen, the others strip off more easily. In this way efficient transport of oxygen is obtained.

The reason lies in the fine structure of the molecule. The five bonds to iron in venous haemoglobin get all their electrons from the nitrogen atoms of the porphyrin and histidine. The iron atom is in a doubly ionized state (it has lost its two outermost electrons) and the six electrons in the next outermost layer are in paired orbitals of their own. One of these pairs is

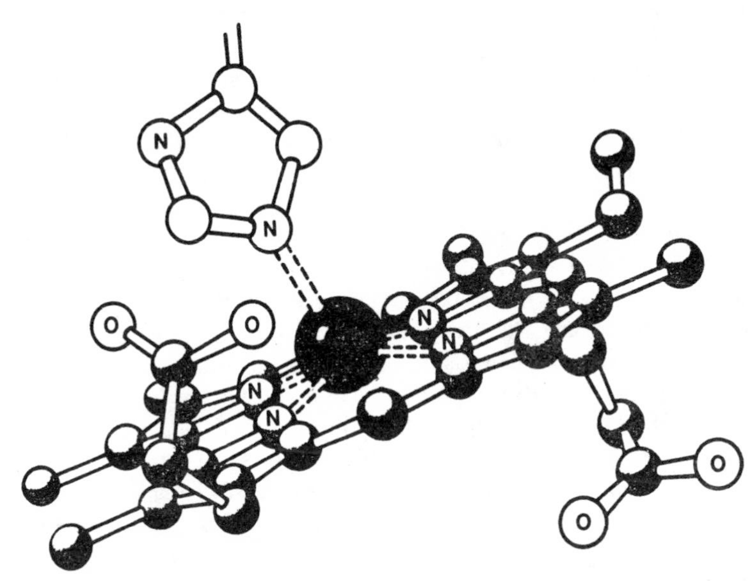


FIGURE 13 The iron atom at the active centre of haemoglobin is joined to the porphyrin ring and to the nitrogen atom of a histidine residue in the protein (globin) chain. When an oxygen molecule links to the iron atom, the iron orbitals change shape. This, in turn, makes the iron atom move closer to the porphyrin, tugging the globin chain into a different shape – a mechanism which facilitates the combination of oxygen molecules with the other iron atoms in the molecule.

positioned so that the iron is repelled from the plane of the porphyrin ring by neighbouring porphyrin orbitals and lies slightly above it, toward the nitrogen of the histidine residue to which it is bonded. When a sixth coordinate bond is formed to an oxygen molecule, the orbitals move into different positions, and the iron atom now moves into the plane of the porphyrin. The movement is through a distance of about 1 ångström. Although this is a tiny amount, the pull on the histidine residue alters the configuration of the globin chain. This alteration changes the relationship between the different globin chains in such a way that the small constraints which otherwise slightly hinder oxygen attachment are lowered. Consequently, once primed with one oxygen molecule, the haemoglobin is successively more likely to gain the remainder of its complement. It was not until the summer of 1970 that Perutz put the seal to his work on haemoglobin by realizing that this was the molecule's chemical secret.

The story of haemoglobin does not stop there, however. More recently, a number of chemists have been attempting to synthesize iron compounds which will act as models of haemoglobin in their behaviour towards oxygen. During 1974, James Collman and colleagues at Stanford University in California reported the synthesis of a modified iron-porphyrin which would bind oxygen molecules reversibly at room temperature. It is possible that further development of such models could have important therapeutic applications in treating people with abnormal haemoglobins, such as sickle-cell anaemia sufferers.

The strangely symmetrical porphyrin ring is not restricted to haem, but occurs elsewhere. Where metals are involved in biological reactions it is one of nature's favourite structures. The essential dietary substance vitamin $B_{12}$, for example, consists of a porphyrin ring (rather more complex than that in haem) with a cobalt atom in the middle. It was for determining the structure of vitamin $B_{12}$ by X-ray methods that Dorothy Hodgkin, whose name has already been mentioned in connection with the structure of insulin, was awarded her Nobel prize. Porphyrins in which the metal atom is copper, manganese and vanadium have also been found in natural systems.

Incorporation of metals into proteins with the aid of a porphyrin is not the only example of a biological molecule in which the part produced as a direct result of the genetic code – the protein – is not sufficient by itself to produce the desired effect. Vitamin $B_{12}$ has been implicated in the activity of several enzymes, and it is now becoming apparent that many of the vitamins – substances identified, for the most part, in the early years of this century and found to be essential to healthy life – are necessary to make enzymes work properly, and are then called cofactors or coenzymes.

Other enzymes contain just a metal atom. For example, carboxypeptidase A, an enzyme which chops the end amino acid from peptide and protein chains, requires a single atom of zinc to work. Study of these complex systems is still at a very early stage, and it is not yet possible to generalize about them. However, it seems likely in some cases that the basic function performed by the biological system can be performed by the metal ion or organic cofactor alone, but only with a low efficiency. The association with protein, followed by subsequent evolution of the protein chain, has led to much enhanced efficiency and made the continuance and development of life possible. The reverse can also happen: the metal may also help the protein. Thermolysin, for example, requires one atom of zinc per molecule for its activity. It also contains several calcium atoms per molecule. Unlike most enzymes, thermolysin is stable at temperatures near the boiling-point of water. X-ray analysis of the structure has shown that this stability is a result of the association with calcium ions. If the ions are removed, the enzyme is still reactive, but is no longer resistant to heat degradation.

## Let there be light

One complex system, which involves no protein, but does involve a metal atom and the porphyrin ring and which may, more than any other, have underlain the continuance and development of life, is chlorophyll. This substance plays a key role in photosynthesis, the process which mediates the basic conversion of inorganic into organic materials – or, since these old divisions are now crumbling, the building up of carbon chains from single carbon molecules. Despite the cleverness of the genetic code, no animal can survive without the supply of some preformed carbon chains. Ultimately, it is to the plant kingdom that we owe our lives.

Although a few types of bacteria have developed life styles which allow them to survive on a diet of mineral salts and other inorganic chemicals, all higher life forms depend on the photosynthetic activities of plants. In overall terms, green plants absorb light and use its energy to convert water and carbon dioxide into carbohydrates – biochemicals composed solely of carbon, hydrogen, and oxygen. The primary step in photosynthesis is the absorption of light energy; the key compound in this is chlorophyll a, a large part of which is a porphyrin ring with a magnesium atom in the middle. When this absorbs light, an electron is excited into a higher orbital. There are several other compounds in green plants which can absorb light but in all cases, the really important molecule is chlorophyll a, for when the other molecules absorb light, they pass the excitation energy to an adjacent chlorophyll a molecule before it can be used further.

Before considering the next step, it is worth mentioning that we can be sure we know the correct structure of chlorophyll a, because it has been synthesized in the laboratory. Despite the sophistication of analytical methods now at their disposal, chemists still like, if possible, to make artificially the structure they have elucidated, just as a double-check on the accuracy of analysis. Chlorophyll a was synthesized in 1960 by Robert Burns Woodward of Harvard University. One of the most prolific synthetic chemists of our time, Woodward, together with large teams of research workers, has succeeded in synthesizing many biological chemicals of complex structure, including vitamin $B_{12}$. For his achievements Woodward was awarded the 1965 Nobel prize in chemistry.

The energy trapped in an excited molecule of chlorophyll a can escape again as light. Under certain conditions chlorophyll a fluoresces. Whenever a molecular electron is excited to a higher orbital, some of the excitation energy is always dissipated as heat, so that fluorescence occurs at a longer wavelength (lower energy) than that of the light initially absorbed. The other photosynthetic pigments all absorb at shorter wavelengths than chlorophyll a, so that even when some of the energy is lost, they can still excite one of the latter's electrons. It is possible that these pigments have only evolved so that plants and algae can use a broader part of the visible spectrum than would be available if there were only one photosynthetic pigment; however, absorption is still quite restricted, as witness the almost universal greenness of plants.

Assuming that the chlorophyll a, instead of re-emitting its energy as fluorescence, disposes of it fruitfully, what happens? There is, at the moment, no completely clear answer. It is known that the overall process involves two light-absorption steps, each of which results in the freeing of an electron (derived ultimately from the breakdown of water into gaseous oxygen and hydrogen ions). The electrons are carried down complex gradients, involving a number of other biochemicals, such as the cytochromes, in such a way that two crucial substances are formed. These are adenosine triphosphate (ATP) and reduced nicotinamide adenine dinucleotide phosphate (NADPH).

ATP, formed by this mechanism from adenosine diphosphate (ADP) and inorganic phosphate, is the universal energy currency of life. It plays a role, for example, in the manufacture of proteins by converting amino acids to aminoacyl phosphates, which are more reactive. It is the modern biological equivalent of the phosphates believed to have played a part in the condensation reactions on the primeval Earth, but its advantage over them is that it recycles easily. When it forms an activated phosphate compound, ADP is produced as a side product, and, as in the case of photosynthesis, there are biological mechanisms which utilize energy from other sources to reconvert ADP to ATP. NADPH, on the other hand, is used for storing hydrogen atoms until they are required in biological reactions, when it gives them up to become NADP, the substance from which NADPH is made during photosynthesis.

This seems to be a very complex process for storing the energy from light but, in many ways, it is necessary. Electron excitation is a quick way of grabbing energy as it passes through a cell in the form of light. By its nature, an excited electron is in an unstable state and wants to get rid of its excess energy. An excited chlorophyll a molecule has about $10^{-9}$ seconds in which this can happen in ways other than the wasteful form of fluorescence. Consequently, a system has had to evolve which can turn transiently held energy into a more permanent form in which it can be used at leisure. Once the ATP and NADPH have formed, the remaining steps in photosynthesis, the so-called Calvin cycle, can take place in the dark.

The working out of the actual chemistry of carbon-compound synthesis in the photosynthetic scheme was the result of the bringing together of two at that time new techniques, and their dedicated application by a group at the University of California led by Melvin Calvin (who has also played a significant part in chemical origins of life studies).

Shortly before the beginning of World War II, Samuel Ruben and Martin Kamen began to study photosynthesis at the University of California, using newly discovered isotopes of common elements. Through the use of $^{18}O$ they established that the oxygen given off by photosynthesizing plants comes from water, and not from carbon dioxide. When this heavy isotope of ordinary oxygen was incorporated in the water used in their experiments, it could later be detected in evolved oxygen, whereas when they incorporated it into carbon dioxide none of the isotope appeared in the oxygen. They then extended their studies by using a radioactive isotope of carbon ($^{14}C$). However, the advent of war, and the subsequent death of Ruben in a laboratory accident, cut short their work. By the time the war had finished, the technology of isotope separation and purification had advanced greatly as a by-product of the development of the atomic bomb, and $^{14}C$ was much more readily available. In 1946 Calvin organized a group to trace the path of carbon in photosynthesis. In addition to a better supply of $^{14}C$, he had another asset, the analytical method of paper chromatography, developed during the war by Martin and Synge.

The basis of Calvin's tracer work was similar to that of Ruben and Kamen with $^{18}O$. Feed an unusual isotope of a common element into a plant, find out which compounds that isotope gets into, and then you can say that the biosynthetic mechanism of the plant involves the labelled atom you have used in making that compound. Where the labelled atom is carbon, analysing mixtures of compounds obtained from the plant at different times after feeding the labelled compound to it makes it possible to establish the order in which the plant synthesizes those compounds bearing the label. The analysis consists of measuring the quantities of isotope in each compound.

One of the first points that Calvin and his colleagues discovered was that photosynthesis worked too fast to make the use of leafy plants feasible in the early experiments. Consequently, most of Calvin's work was with two photosynthetic single-celled organisms, the algae Chlorella pyrenoidosa and Scenedesmus obliquus. These organisms were grown in solution; at a precise time, radioactive bicarbonate ion (which can be used as a $CO_2$ source) was injected into the solution. After a specified period the cells were killed, and extracts made and analysed by chromatography and autoradiography. In order to obtain a meaningful pattern of results, it was necessary in some experiments to kill the algae within a few seconds of exposure to labelled material, so quickly did they use it. Within 30 seconds, the algae had incorporated radioactivity into between twenty and thirty different compounds.

Basically, in the Calvin cycle as it is now understood, a molecule of carbon dioxide forms a carbon-carbon bond with a five-carbon chain compound, ribulose-1, 5-diphosphate; the resultant six-carbon compound splits into two three-carbon molecules. Through the action of ATP, NADPH, and various enzymes, some of these three-carbon molecules are built up to form six-carbon molecules, which then react with another three-carbon molecule to form a five-carbon and a four-carbon molecule. The latter then gains another three-carbon chain to form a seven-carbon compound, sedoheptulose-7-phosphate. This reacts together with another three-carbon molecule to produce two five-carbon molecules which are converted into ribulose-1,5-diphosphate, and the cycle begins again.

By other biosynthetic pathways – many of which have been studied using isotope tracer techniques – the carbohydrates formed can be transformed into other carbohydrates and carbohydrate polymers such as starch, into amino acids, fatty acids (porphyrins are probably biosynthesized from the amino acid glycine and the simple analogue of the fatty acids, acetic acid), and ultimately all of the chemicals of life.

Through the absorption of light, plants make all higher life forms possible, including man who, seeing what the plants do, has struggled to achieve an understanding of how they do it. Sight is possibly the most important of the senses to man and other higher animals. What does that part of the electromagnetic spectrum which we call visible light do to enable us to see?

Such a question involves an understanding of the human brain which far transcends not only modern chemistry, but all of modern science. But, just as chlorophyll acts as a mediator between untrapped light energy and the firmly held energy of a chemical bond, so are there chemicals and chemical reactions at the interface between the photons of light that strike our eyes and the electrical impulses which pass from the optic nerve into the brain for analysis. Not surprisingly, the chemistry of that interface obeys the simple chemical rules elucidated in earlier chapters and, like much of this chapter, helps to show how the manipulation of molecular shapes can produce effects of great diversity and sophistication.

Physiologists have shown that the human brain, under carefully controlled conditions, is capable of registering a sensation of light when no more than two or three individual photons strike the eye. It has even been suggested that a single photon striking the eye and causing a reaction in one molecule is sufficient to be registered.

During the 1930s, Professor George Wald of Harvard University, at that time working in Berlin and Zurich, discovered that the molecule in question, the visual pigment rhodopsin (also known as visual purple), which had been found in the retinal rod cells of many types of animals before the end of the nineteenth century, consisted of two parts: a colourless protein (opsin) and a previously unknown substance, now called retinal, one of a class of compounds known as carotenoids.

Toward the mid 1940s, R. A. Morton and his colleagues at the University of Liverpool showed that there was a close chemical relationship between vitamin A, an essential dietary ingredient the lack of which causes a number of bodily disturbances, including 'night blindness', and retinal. He showed that retinal has two fewer hydrogen atoms in its molecule than vitamin A (they are removed by the conversion of NAD, a close relative of NADP, to NADH). There are also close chemical similarities between both these compounds and the carotenes widely found in plants. It is believed that vitamin A is produced in the body by splitting carotene molecules in half.

In addition to the differences between the numbers of hydrogen atoms in retinal and vitamin A, there is another important difference. These molecules are characterized by systems of alternating single and double bonds. Double bonds between carbon atoms can produce different compounds made up from the same ordering of atoms in the molecule. Known as geometrical isomers, these molecules differ only in their three-dimensional structures. The structures would be identical if rotation around one of the double bonds could take place. However, the effect of the second bond in a double bond is to increase the rigidity of the molecule, and this prevents such rotation (see Figure 7, p. 47). Chemists have a convention for describing double bonds as either cis or trans, according to exactly how the remaining groups attached to a pair of doubly bonded carbon atoms are situated in space. All the double bonds in vitamin A are trans; in retinal, one of them is cis. This single cis bond gives the molecule an uncomfortable shape. It is under strain which makes the molecule less stable than it would be if, like vitamin A, the double bonds were all trans. This, as workers in George Wald's laboratory at Harvard found, is the secret of vision.

In 1952, Ruth Hubbard showed that only the cis isomer of retinal would combine with the protein opsin to form the visual pigment rhodopsin. Seven years later it was discovered that the effect of light on rhodopsin is to excite an electron in retinal out of its bonding orbital. The electron excited is one of those involved in the cis double bond. Freed momentarily from the constraint of this bond, the retinal molecule changes shape, so that when the electron falls back into a bonding orbital reforming the double bond, it locks the molecule into the lower energy configuration of trans retinal. This new shape does not suit the protein opsin, and the two molecules separate. As Ruth Hubbard wrote in 1967: 'Everything else – further chemical changes, nerve excitation, perception of light, behavioural responses – are consequences of this single photochemical act.'[6]

Since then, there has been a considerable amount of research into the chemistry of vision, to gain a more detailed understanding of the processes involved. For example, it is known that photolysis of a single rhodopsin molecule causes the release from the surrounding membrane of up to 1,000 calcium ions. These probably inhibit the normal 'dark current' of sodium ions, thus triggering a nerve impulse. But how the amplification effect is achieved is still not known.

It is also known that, between a photon hitting a rhodopsin molecule and the appearance of separate products, there are a number of very short-lived stages, some of which are reversible. And the irreversibility of later stages may be due as much to changes in the shape of the protein – through alterations of the sulphur-sulphur bridges – as in the retinal molecule. As further information becomes available, through the improvement of techniques for studying very fast reactions, a deeper and fuller understanding of the visual process should emerge.

### A matter of taste (and smell)

The geometry of chemical molecules thus plays an important part in such diverse functions as oxygen transport in the blood, catalysis by enzymes, and vision. What of the other senses? Two at least, taste and smell, involve direct contact with and analysis of a wide range of chemical substances, although the analysis is in the subjective form of 'pleasant', 'unpleasant', 'bitter', 'burnt', 'sweet' and 'fruity', rather than in the objective chemical sense of structure analysis. To produce even a limited range of distinct sensations, it seems likely that substances must react chemically with receptors in the mouth and nose as the first step toward producing the nervous impulses that lead to our brains.

Many small carbohydrate molecules, collectively called sugars, taste sweet. One of the most effective of these is sucrose, which is what supermarkets sell as 'sugar'. All the sugars belong to the same family of com-

6. Ruth Hubbard and Allen Kropf, 'Molecular isomers in vision', *Scientific American*, ccxvi, 6, 1967, p.64.

pounds and have common structural features, but there are other substances, chemically quite unlike sugars, which also taste sweet. Saccharine is an example. To account for the diversity of chemical structure and the common sensation of sweetness. Professor R. S. Shallenberger of the New York State Agricultural Experimental Institute suggested in the mid 1960s that sweetness resulted from the formation of hydrogen bonds between the sweet molecules and receptor substances in the tongue. As part of his early work, which concentrated on the sugars themselves, Shallenberger pointed out that the simple sugars that are less sweet have more internal hydrogen bonds – that is, the bonds form between different atoms within the molecule – than those which are more sweet. Possibly, he suggested, sweetness depends on the number of hydrogen bonds which can form with receptors in the tongue.

In 1967 Shallenberger extended his theory. By this time he had found structural correlations among a wide range of different sweet-tasting chemicals. All of the compounds he investigated not only had a hydrogen atom in one part of the molecule available for hydrogen bonding but, at a distance of not less than 2·5 and not more than 4 ångströms from this, was another atom able to form another hydrogen bond. Shallenberger suggested that the receptor molecule must contain a similar two-pronged unit, enabling the sweet molecule to combine with it by forming two hydrogen bonds.

An interesting chemical accident in 1969 helped to support Shallenberger's theory that the sweetness and the complementary receptor portions of molecules must consist of at least two parts. During work on a peptide hormone (gastrin), Robert Mazur, James Shlatter and Arthur Goldkamp, chemists working at the Chemical Research Department of the G. D. Searle Company in Skokie, Illinois, discovered a dipeptide that was 200 times sweeter than common sugar. One amino acid, alanine, was already known to be sweet, but this is only twice as sweet as sugar, and was not a component of the dipeptide, which was made from aspartic acid and phenylalanine (which, by itself, is bitter rather than sweet).

The Searle team prepared a number of dipeptides. All the others containing phenylalanine were bitter, although several containing aspartic acid were sweet. Work at Searle has continued in this field and, in 1971, Mazur, Goldkamp and Patricia James published a paper analysing the criteria of peptide sweetness. The aspartic acid part is necessary, and both its amino group and the acid group in its side-chain have to be free (presumably to form hydrogen bonds), while the other end of the molecule (the second amino acid) must contain a hydrophobic group – that is, a molecular structure which has no affinity for water. The function of the second part may be to prevent interference by other molecules with the hydrogen bonding process.

At the time of the Searle chemists' original discovery, Takashi Kubota and Isao Kubo of Osaka City University in Japan had just published a paper about bitterness. They showed that a series of bitter-tasting compounds also had the two-pin-socket arrangement found in sweet-tasting compounds, but that the optimum distance of the two pins was 1·5 ångströms apart – much less than for sweetness.

Clearly, delicate chemical geometry is involved in the determination of taste. There is evidence that the receptor molecules are proteins, but before the story proceeds, it will be necessary to know more about the structure of such proteins, to see how they interact with small molecules.

Although only four basic taste sensations (sweet, bitter, salt, sour) are recognized by physiologists, there are about twice that number of different basic smells, all of which come down to a question of basic molecular geometry. The first person on record to have suggested that this might be so was the Greek philosopher Lucretius. No one took him very seriously until 1949, when the Scottish chemist R. W. Moncrieff proposed the existence, in the olfactory apparatus, of a few different types of receptor, each corresponding to a different primary odour. This idea was interpreted in stereochemical terms by John Amoore while he was still an undergraduate at the University of Oxford. During the 1950s and early 1960s, Amoore, in collaboration with James Johnston and Martin Rubin of Georgetown University School of Medicine, set out to test his theory experimentally.

Amoore searched the literature of organic chemistry, comparing the known smells of simple molecules with their general shapes. He came to the conclusion that there are seven primary odours, and in general terms described the probable shape of the receptor site for each. A complex odour, he suggested, results from a molecule fitting more than one receptor shape. This hypothesis was tested by synthesizing an organic molecule which consisted of three chains of carbon atoms attached to a central carbon atom, to which a hydrogen atom was also attached. A fruity smell was predicted for this molecule because it fitted three of the theoretical receptor shapes ('floral', 'pepperminty' and 'ethereal'). When the single hydrogen atom attached to the central carbon was replaced with a methyl ($CH_3$) group, the overall shape of the molecule was changed so that it should have fitted less well into the hypothetical 'floral' and 'pepperminty' receptors, but just as well into the 'ethereal' receptor. When tests were carried out on humans, it was found that the first compound did have a fruity odour, while in the second an ether-like tinge was clearly detectable.

Using bees, which 'smell' with their antennae, as well as human subjects, Amoore, Johnston and Rubin went on to show that molecules which fit the same single theoretical receptor site all have the same smell, while two molecules, each shaped to fit a single, different site, always smell different. Physiological evidence in favour of the theory came from another source when R. C. Gesteland, working at MIT, studied the stimulation by different substances of single olfactory nerve cells in the frog. He found that cells responded differently to different chemicals. Since a nerve impulse is always a yes/no situation – a cell either responds or it does not – this experiment appeared to confirm that there are different receptors attached to the business end of the nerve cells. Gesteland estimated that there are probably eight different types, at least five of which correspond to the primary odours suggested by Amoore. However, more recent work by French scientists suggests that most odour receptors may not react specifically to one type of odour. Instead, they believe that different receptors show different levels of sensitivity to particular odours, and that the sensation of smell results from the pattern and intensity of response of a large area of receptor cells.

In man, smell is an important part of life, but it is not the essence of life and death as it is to some other species. Without smell, some of these would find it difficult to survive, for the finding of mating partners is directly connected with this sense. Here the word 'smell' is being stretched a little, for, as with the bee, detection of the 'odorous' substances may not be through what we would call a nose.

One particular form of chemical detection is of prime importance. Some species produce a chemical which has a strong effect on other members of the same species, when it is 'sniffed' by them – similar in some ways to the effects of hormones. Since 1959 such communicating chemicals have been called pheromones.

Important pheromones – not only to the species that use them, but also of potential value to man – are the insect sex attractants. Female insects of species such as the silkworm and gipsy moth release small quantities of pheromone as an advertisement, so that males in the area can find them. It has been calculated that the average female gipsy moth carries enough sex attractant to communicate with more than $10^9$ male moths. Even though the total amount is tiny, only a few molecules are needed to stimulate a male moth into looking for the source of the smell.

The smallness of the total amount of active material in any single moth is shown by details of the isolation by chemists of two such substances. Adolf Butenandt and his colleagues, working at the Max Planck Institute of Biochemistry, Munich, Germany, extracted material from a quarter of a million silkworm moths in order to obtain twelve-thousandths of a gramme of the sex attractant bombykol, while U.S. Department of Agriculture scientists Martin Jacobson, Morton Beroza and William Jones used twice that many gipsy moths to get twenty-thousandths of a gramme of their sex attractant. Both compounds were identified and synthesized in the late 1950s and early 1960s, since which time it has become clear that many other insects – and higher animals – use pheromones as part of their intercommunication systems. Certain species of ants, for example, provide chemical route markers to show others in their colony the way to sources of food, and produce other pheromones to warn colleagues of any danger they have encountered.

In recent years, there have been literally hundreds of scientific papers written about the different chemicals used as pheromones by insects – and other species, for it has been discovered that some sea anemones have an 'alarm' pheromone, while even seaweeds use sex attractants – and it is now clear that the situation is highly complex. In some species, a pheromone may be more than one substance, with the components needing to be mixed in particular proportions to be effective. A more interesting aspect of the subject, only just beginning to be explored, is how species developed their pheromones. Some seem to use plant substances, which they concentrate or modify slightly. Oak-leaf roller moths, for example, appear to extract their sex stimulant pheromone from oak leaves – a discovery that arose from observations of male moths trying to mate with the leaves! The validity of this research has, however, been questioned by some scientists, so the origin of pheromones is still an open question.

The importance of these substances to man is primarily that they offer a way of controlling specific insect pests. Occasionally there is a secondary commercial value. In 1957 Butenandt isolated a simple substance produced by male specimens of a tropical water bug to excite females sexually. An extract of this substance, which to humans has an odour like cinnamon, has been used for many years in south-east Asia as a spice. Now the synthetic substance is sold instead. However, this use is trivial compared to the primary one mentioned above.

Insect pests cost mankind millions of pounds each year in terms of damaged crops. Yet, as is made clear in Chapter 7 man's attempts to combat these pests chemically have, until recently, met with mixed success. Synthetic pesticides have been insufficiently selective and, as a consequence, have caused serious pollution problems. In addition, insects are adept at developing strains resistant to a particular pesticide.

Combining the techniques of chemistry with an insect's own characteristic biology provides the possibility of selective attack. No insect can develop a resistance to its own sex attractant – if it did, the species would soon die out – and, because the insects have been efficient in developing attractants that attract only their own species, other species are unlikely to be affected by insect destruction methods that involve a particular pheromone. However, it was found in mid 1972 that certain Indian termites and honeybees share a common pheromone. And, early in 1973, U.S. scientists reported that cabbage looper moths and alfalfa looper moths have the same sex attractant; they do not interbreed for behavioural reasons.

When the U.S. Department of Agriculture scientists had identified gipsy moth sex attractant, they were able to synthesize it in the laboratory. They also synthesized some very similar compounds, one with two fewer carbon atoms (which had no effect) and one with two more (which was just as effective as the natural material). As this latter product (gyplure) could be synthesized very easily from a substance that occurs naturally in castor oil it was adopted by the nation for a programme of attack on the gipsy moth. Traps were baited with gyplure and the moths flew straight into them, after which they could be easily destroyed.

Although the use of pheromones is an excellent way of attracting insects to a particular place, it does not kill them. There is one variation that goes all the way. It is possible to use some pheromonal substances to encourage females to lay eggs in unsuitable places where they are unlikely to survive. But, in general, it is necessary to kill the insect by a second method. However, the principle of using a particular species' own biological chemistry to destroy it can be extended.

Insects have hormonal systems for communication within a single organism, and those can be turned against them. There are two distinct phases of the insect life cycle, a larval stage and an adult stage. Since the existence of insect hormones was established in 1934 by British biologist Sir Vincent Wigglesworth, it has become clear that three hormones are frequently involved in the change from larva to adult. There is a brain hormone which stimulates production of a moulting hormone (ecdysone), which is responsible for inducing the shedding of the larval skin when the protoinsect inside gets too big for it. Ecdysone works in conjunction with juvenile hormone, which keeps the insect in the larval stage. The laboratory syntheses of both juvenile hormone and ecdysone were achieved in the mid 1960s, following work similar in complexity to that required for isolation and structural elucidation of sex pheromones. For example, a close relative of insect moulting hormone, crustecdysone, was isolated in 1966; two-thousandths of a gramme were obtained from a ton of crayfish. Incidentally, it was discovered in 1972 that several species of crab use crustecdysone as a moulting hormone and a sex attractant. A male crab, attracted to a female by the crustecdysone, protects her while she moults, then mates with her.

If the manufacture of juvenile hormone is not switched off at the right time, instead of metamorphosing into an adult a larval insect continues to moult – perhaps once or twice more than it usually would – and then dies. As they are unable to take part in the reproductive process until after metamorphosis, insects that are kept 'juvenile' do not produce offspring.

The potential insecticidal properties of insect hormones were realized by Carroll Williams at Harvard University in the mid 1950s, when he discovered that the Cecropia moth was a rich source of juvenile hormone. It was not until the 1960s that syntheses of the actual hormone substances were achieved by several groups using different synthetic methods. However, by chance, Williams and a collaborator, John Law, discovered that simpler, quite easily synthesized substances had juvenile hormone-like activity. This synthetic juvenile hormone was found to be useful in controlling the species of mosquito which is responsible for spreading yellow fever, and also the body louse, which can spread several human diseases, including epidemic typhus.

Czechoslovak workers led by Karel Slama manufactured similar compounds and found them to be highly active against the bug, Pyrrhocoris apterus. So effective was the material they produced that treatment of an adult male with it rendered the eggs of any female with which the male copulated sterile for the rest of her life.

In an unexpected way, the work by Williams in America and Slama in Czechoslovakia also showed the importance of international collaboration in science. In the mid 1960s, Slama sent Williams some specimens of Pyrrhocoris apterus for experimental use. However, Williams could not get them to metamorphose properly, although Slama had bred thousands of adults of the species and Williams used his breeding methods.

Williams checked all the possibilities to make sure that there was no difference between his breeding conditions and Slama's. Eventually, after eliminating all other possibilities, only one remained. Part of the procedure involved small strips of paper being put into the glass breeding tanks for the insects to exercise on. Williams replaced the paper he had been using with a different type. The bugs began to breed properly.

American paper is frequently made with pulp from the balsam fir, while European paper is not. The balsam fir contains a substance, juvabione, which has subsequently been identified and synthesized, that acts on Pyrrhocoris apterus as if it were juvenile hormone.

There is a moral to this story. After some years of using non-specific synthetic insecticides, man concluded that it might be better to look for specific defence systems against insects. This might seem to be a real example of human ingenuity, except that certain plants have been doing it for years. In addition to balsam fir, it has been known for some time that the Australian timber tree, Podocarpus elatus, is particularly resistant to insect attack. In 1966 it was discovered that the tree is a rich source of crustecdysone, which interferes with proper insect development.

In the early 1960s it was shown that one of the intermediate compounds formed during the biosynthesis of ecdysone is cholesterol, a substance for which at least some insects have a dietary requirement. Cholesterol also occurs in human beings, and is structurally related to the steroid hormones, which control various aspects of sexual activity. The economy of nature is apparent: certain basic chemical structures are used over and over again, and it is by small changes in parts of the molecule, which alter its geometry and thus its reactivity, that the diverse activities of biological systems are mediated.

Underlying the superficial diversity of living nature there appears to be a unity based on a few simple chemical principles. If, for the words 'living nature' we substituted 'chemical industry', the sentence still reads true. Although it would be possible to fill the remainder of this book with further details of the ways in which chemists are laying bare living processes, it is now time to look at chemistry from a different angle. From Homo sapiens, man the thinker, we pass to Homo faber, man the maker.

# *CHAPTER 5*

## INDUSTRIOUS CHEMISTS

The origin of life, or of the molecules that were necessary for formation of life, followed basic chemical laws, such as those of kinetics and thermodynamics. The time scales involved were enormous, because the processes were largely random. In the laboratory experiments that simulated the primeval production of such important building blocks as amino acids and sugars only a few days were needed to repeat nature's long-term experiments.

One of the characteristics of laboratory chemistry is that it seeks to increase the yields of reaction and produce as much of a desired product as possible. Production of chemicals on an industrial scale might be expected to follow the same search for efficiency, but this is not so. Although industry must be efficient if it is to make a profit, the yields of many of the reactions used seem to parallel those postulated for the origin of life, rather than the practices of the laboratory chemist. Yet clearly this is a successful procedure, for chemical manufacture is one of the world's largest industries.

Chemical industry is called a service industry, for its primary function is to serve the requirements of other industries, many of them consumer industries. A consumer industry produces artefacts for sale to the general public – cars, television sets, clothes, and so on. Only about a quarter of the chemical industry's products are sold direct to the consumer; detergents are an example. The rest go to other industries as raw materials for textile fibres, for dyeing, or fabricating machine parts, or for use in such processes as steelmaking, which requires large quantities of strong acids.

The chemical industry has been an important part of society since the early nineteenth century. However, the chemical industry in the last century bore little resemblance to today's industry which, in the U.S.A. alone, employs over one million workers and had a turnover of more than $120 billion in 1977.

For its first century the chemical industry followed the pattern of laboratory chemistry, producing new compounds in the same way as they were produced in the laboratory, but on a larger scale. Nowadays this is only true of the production of such low-tonnage materials as pharmaceuticals and dyestuffs. Processes used in factories were taken from the laboratory, scaled up to larger sizes, and then improved on a trial-and-error basis. Today a new chemical process should operate well from the start, because the scale and cost of new plant are so enormous that there is little room for mistakes.

Many of the processes used in the chemical industry involve chemistry that has been known for a long time, and the main change in these areas in recent years has concerned scale of manufacture. But the real heart of the modern chemicals industry lies in petrochemicals, chemicals derived from oil, which lead, among other things, to the production of most of the plastics currently available. Petroleum was only viewed as a source of chemicals in the 1920s; the major growth has been since World War II and, despite the crisis precipitated in late 1973 by the Arab oil embargo, the industry will remain heavily dependent on oil as its major source of raw materials for many years.

Oil companies originally decided that they might utilize some of the materials left over from the manufacture of gasoline, which only takes a small fraction of crude oil. Consequently, the first petrochemical works were sited near oil wells. Since 1945 oil refineries have increasingly been sited near the major points of use, large urban centres throughout the world. This change has led to the growth of a world-wide petrochemical industry, rather than one existing almost exclusively, as was the case before the war, in the United States. The demand for chemical raw materials has become so great that one large British chemical company owns its own chain of petrol stations to get rid of the 'waste product' left over from crude oil after the chemicals needed for its manufacturing processes have been extracted.

In 1938, Leo Baekeland said that 'the whole fabric of modern civilization becomes every day more interwoven with the endless ramifications of applied chemistry.' This is certainly true today, although it is worth remembering that fewer than 1 per cent of the chemical compounds known to man have ever been used industrially. The extent to which chemistry permeates every aspect of our daily lives is discussed in Chapter 6. This chapter describes the petrochemical industry and some of the aspects of how industry works. As has been intimated, industry has one constraint that the research worker does not often feel. In addition to obeying the laws of nature, it must also obey the laws of economics.

## Taking economics into account

To combine the laws of nature and the laws of economics into a simple example may give some insight into the way in which the industrialist thinks differently from the research chemist. If we take a simple reaction in which compound *A* is converted into compound *B*, and a study of the kinetics shows that this a simple half-life reaction (see note, p. 61), the amount of *B* produced depends in part on the amount of *A* used. In a certain amount of time – the half-life – half the amount of *A* present will have turned into B. After the same amount of time further, half the remaining *A* will have turned into *B*. If the half-life is 30 minutes, after an hour the reaction mix will contain a 75 per cent yield of *B* and 25 per cent unconverted *A*.

Imagine that this reaction is best performed in a boiling water solution. If a research chemist undertakes the reaction because he wants to study some property of compound *B*, he will note from early experiments that the half-life is about half an hour. Since he wants as much *B* as he can get, he may start the reaction before he goes to lunch and stop it after he gets back, letting it run for perhaps two or two and a half hours. According to the half-life rule, he will then have a yield of about 94 or 97 per cent of *B*, which, from his point of view, is very good.

From the perspective of the industrial chemist, who perhaps wants to produce 10 tons of *B* to sell, this method of working is disastrous. If he lets the reaction run for two and a half hours, the last half hour costs him as much to produce 3 per cent of *B* as it did to produce 50 per cent in the first half hour. He is paying the same amount each hour for supplying energy to the reaction to keep it boiling, for paying the workers who look after the plant, and so on. Clearly, he must take into account more factors than the laboratory chemist.

For example, in some cases it will be worth his while to spend that extra money just to get the yield of *B* up to 97 per cent. If it is very difficult and consequently costly to separate *A* from *B*, it may even be worth letting the reaction run for three hours and then selling the 98+ per cent pure *B* as it is, without removing the traces of unreacted *A*. On the other hand, if the customer wants 99·9 per cent pure *B* (this is not unlikely) it may be more economical to use an expensive separation process than to keep the reaction going for five hours.

At the other extreme, if *B* separates easily from *A*, by filtration for example, it will probably be wiser to run the reactor for short times, separate the two materials, and mix the recovered *A* with fresh *A* and repeat the reaction. The modern approach to such recycling operations is to use continuous reactors in which product is continually removed and fresh material added at the same rate.

This is a simple example of the type of economic constraint felt by the industrial chemist. Another example might involve the temperature at which a reaction is carried out. The rate at which reactions occur depends on temperature, and a drop in temperature in the lab may mean that an experiment takes half an hour instead of 25 minutes. On an industrial scale, this would mean an increased running cost of over 15 per cent for the production of that batch of material.

Consideration of these types of problems, and working out the best compromise, is called 'optimization', a procedure that has led to a petrochemical industry in which yields are often small by laboratory standards, and reactions usually give many products rather than one or two. Optimization also includes taking into account how those different products will be used. Not only does a by-product that will not sell have no value, it frequently has negative value, such as the cost of disposing of it without polluting air, land, or water.

The approach used in the chemical industry is called 'unit operations', an engineering concept which has been in use for nearly a century. A process is broken down into constituent parts or operations, each of which is looked at as a unit. Thus, a very simple process might involve only two unit operations, reaction in the reactor and separation after the reaction.

To the chemist, the heart of an industrial process will always be the reactor. Early reactors developed from the type used in the laboratory – they were little more than overgrown flasks. But today many of the products of chemical industry go into reactor-building: special steels, special glasses for lining metal reactors. Today's chemical reactors are designed to withstand rigorous conditions and to keep on working without fail, for failure of any part in a modern large-scale chemical plant can cause a major disaster in both human and financial terms. When a temporary pipe fractured at the Nypro chemical plant in Northern England in June 1974, the explosion caused by release of cyclohexane vapour killed twenty-eight people and caused millions of dollars worth of damage.

As a reaction is scaled up from lab bench to factory, an important mathematical change has to be taken into account. If one thinks of a cylindrical reactor, it is easy to calculate how much it will hold and what its surface area will be. With a cylinder twice the height and twice the diameter of the first, the surface area will be roughtly four times as great, but the volume increases eight times. For a reaction which produces heat, in which temperature has to be kept constant, the heat must escape through the walls of the larger reactor twice as fast as through the walls of the smaller reactor. This may mean that while the first reactor loses heat from the reaction to the surrounding air, the second will need to be designed for more effective heat transfer.

This may seem to make large reactors a bad proposition, but possibly some other operations in a process will require heat. If the heat given off by the reaction is used to generate steam a reaction may be kept within the temperature limits required and the steam piped to another part of the plant, thus turning what is initially a nuisance into an advantage. This is an example of optimization. To be truly economic, one must never waste one's resources – and energy is a valuable resource.

As the chemical industry grew during the last century, the two commodities taken most for granted were energy and water. Energy could be obtained cheaply from fossil fuels, and water seemed to abound everywhere. Yet now there is grave concern in industrialized countries about future supplies of both these commodities. Consequently, conservation and reuse of both by industry is increasingly important.

From our first example of scaling up it can be seen that using a large reactor can have hidden advantages. There are other, more obvious, advantages. Doubling the size of a reactor does not usually require twice the manpower to operate it, so there are savings in plant running costs. There are also disadvantages to large reactors that are just like containers. These batch reactors mean that a process cannot operate smoothly all the time. Reactors have to be emptied, cleaned and refilled. Filling a 20,000 gallon reactor and then heating the contents to the temperature required for a reaction will take time that might be spent more profitably in making more product. This delay can be avoided with a continuous reactor, and it is to continuous reactors that large-scale chemical industry has turned increasingly since World War II.

The most common type is probably the continuous stirred tank reactor (CSTR). This is very similar to the batch reactor, except that raw materials are added continuously, the contents particularly well stirred, and a mixture of product and raw materials continually drawn off. The CSTR is ideally suited to reactions in which the products can be separated easily from the starting materials, and the latter recycled. It is not so useful for obtaining high yields or conversions of material. To obtain a 99 per cent conversion of $A$ to $B$ requires a continuous stirred tank reactor ninety-nine times bigger than the one required for 50 per cent conversion. The reason for this lies in the kinetics of the reaction. If raw material flows into the reactor at a certain rate, the average time any molecule spends in the reactor (mean residence time) depends on the size of the reactor; the shorter the residence time, the lower the number of molecules that will react to produce product. However, in modern industry this problem has been overcome by use of multistage reactors in which the product/raw material mix from one reactor is passed into a second reactor, and possibly into more in sequence. If the first reactor is achieving 90 per cent conversion, a second reactor of exactly the same size will achieve 90 per cent conversion of the remaining raw material, giving an overall conversion of 99 per cent. As a reactor giving 90 per cent conversion needs to be only nine times larger than one giving 50 per cent conversion, to go from 50 per cent conversion in a single step to 99 per cent in two steps means increasing total reactor size only eighteen times, five times smaller than the single reactor required for 99 per cent conversion.

This chapter has scanned briefly some notions current in the chemical industry demanded by that industry's need to conform not only to the laws of thermodynamics and kinetics but also to the laws of economics. One area where economics must affect industrial procedure is in the cost of raw materials. In fact, the cost of raw materials may be the deciding factor between different chemical routes to the same end-products. Nowhere has this been more clearly shown in recent years than in the petrochemical industry.

## Petrochemicals

The raw materials for the petrochemical industry are natural gas, refinery gases, which come from the 'cracking' of crude oil, and liquid hydrocarbons. Although some natural gas is a nearly pure chemical (methane), these raw materials are mostly mixtures from which are obtained the actual building blocks of the chemical industry, secondary raw materials such as methane, ethylene, propylene, butadiene, and higher paraffins and alkenes. At present the percentage of petroleum products used in petrochemical production is about 8 per cent of the total petroleum and natural gas production. It has been predicted that this percentage may double by the end of the century, as the major alternative use – burning the materials as a source of energy – is an appalling waste of a diversity of raw materials. However, in view of the escalation in oil prices since 1973, it is possible that other sources of raw materials may become increasingly important to the chemical industry and radically alter such predictions.

Although the percentage figures sound low, it is interesting to compare the growth of the petrochemical industry as part of the total chemical industry. In the U.S. in 1940, just over eleven million tons of chemicals were produced, 10 per cent of them from petroleum sources. By 1960 the total chemicals produced had risen to 82 million tons, 30 per cent of them of petroleum origin, and by 1970, 41 per cent of the 170 million tons of chemicals produced were petroleum based. Thus, in the thirty years since the start of World War II, the output of petroleum chemicals increased more than sixtyfold. About 95 per cent of the organic chemicals produced in the U.S. are petroleum derived, compared to a world figure of 90 per cent. As there are few inorganic chemicals that can come from petroleum – sulphur, which is often found as a contaminant of natural gas and crude oil, helium, and hydrogen are the major exceptions – this shows the great extent to which petrochemicals have cornered their possible market.

When crude oils are cracked by passage through heated towers in the presence of catalysts, the molecules, which are mainly composed of hydrogen and carbon – the hydrocarbons – break down into smaller hydrocarbons. Refinery gas, obtained by distilling cracked crude, accounts for 15–20 per cent by weight of the cracked material. Typically, refinery gas is made of a mixture of saturated hydrocarbons, containing from one to four carbon atoms: methane, ethane, propane, butanes, unsaturated hydrocarbons – that is, those which contain double bonds (ethylene, propylene, butenes) – and hydrogen.

Industrially the most important of the refinery gases are the unsaturated hydrocarbons, the alkenes. These are more reactive than the saturated hydrocarbons because of the double bonds in their structures. Probably the most important organic chemical in the world today is ethylene, not only as a source of the polymer polythene, or polyethylene as it was originally called, but also of many other chemicals formerly made from acetylene. Like ethylene, acetylene contains two linked carbon atoms, but has only one hydrogen attached to each carbon, while ethylene has two. Although this makes acetylene even more reactive than ethylene, economics have ensured that acetylene has been replaced by ethylene as the preferred raw material for manufacturing a number of important organic chemicals.

By a process known as hydrocracking, in which thermal cracking is performed in the presence of hydrogen, it is possible to produce a refinery gas that consists mainly of saturated hydrocarbons. However, as demand for the unsaturates, ethylene and propylene, frequently outruns supply, it is more common to find ethane and propane converted into these compounds, rather than the other way around.

Thermal cracking of ethane, for example, produces a cracker gas that contains mainly ethylene, unchanged ethane (which can be recycled through the cracker) and hydrogen, large quantities of which are needed for synthetic ammonia manufacture, together with about a dozen other substances in small quantities. As cracking is a process which vigorously tears pieces off molecules, thermal cracking of propane is not so efficient as a source of propylene. Propane cracker gas contains about twice as much ethylene as propylene. The precise ratios of products are affected by the cracking temperature and the time the gas spends in the cracker, so some optimization of product according to need is possible.

In 1960 western European production of ethylene was about 2.5 million tons, a figure which had increased 5-fold by 1977. Nearly half the ethylene is consumed in the manufacture of polythene. The remainder is used for making products such as ethanol, acetaldehyde, ethylene dichloride, ethylene oxide, vinyl acetate, vinyl chloride and ethyl chloride. Ethylene is now such an important raw material for heavy organic chemicals manufacture that in some places it is piped over large distances to different manufacturing sites, and it has been suggested that within a few years it may be regularly piped into chemical plants with such other services as water and electricity.

The major revolution in industrial ethylene chemistry came in the 1950s, when the price of ethylene dropped as a result of big expansion in the petrochemical industry. Since then it has knocked the bottom out of the market for some other chemicals in manufacturing, such as acetylene.

Acetylene is manufactured in two ways. In the cracking of methane at high temperatures, three hydrogen atoms are torn from each methane molecule, and the CH fragments formed join up in pairs to make acetylene ($HC \equiv CH$). Acetylene is still obtained also by an old process, the breakdown of calcium carbide, but this is rapidly diminishing in scale. However, this process, which uses coal as a raw material, may regain some of its former prominence if natural gas prices rise substantially.

One primary use for acetylene until the early 1960s was for vinyl chloride manufacture, the monomer from which the plastic polyvinyl chloride is made. This is a simple manufacturing process in which a molecule of hydrogen chloride joins to the acetylene molecule to give the desired product. Since vinyl chloride is itself unsaturated – it is like an ethylene molecule, except that one hydrogen atom has been replaced by chlorine – the triple bond of acetylene makes the addition process simple.

If hydrogen chloride is added to ethylene, the product is a saturated molecule, ethyl chloride, which will not polymerize. Consequently, a different type of process, called oxychlorination, is needed to make vinyl chloride from ethylene. Ethylene is reacted with hydrogen chloride and oxygen. Instead of the hydrogen chloride adding to the ethylene, the hydrogen part adds to the oxygen to form water, and two atoms of chlorine add to the ethylene to make dichloroethane, which can be cracked to produce vinyl chloride and half as much hydrogen chloride as was originally used. The latter is reused to oxychlorinate more ethylene. Once this process was worked out, it was no longer economic to make vinyl chloride from the more expensive raw material, acetylene.

One of the oldest industrial organic chemicals is ethanol, originally made by yeast fermentation. This method is still used to produce drinking alcohol – whisky, gin, and so on – and in some underdeveloped countries, lacking a petrochemical industry, to supply industrial alcohol.

In the more developed countries nearly all ethanol, which is widely used as an industrial solvent as well as for the manufacture of large-tonnage organic chemicals such as ethyl acetate and acetaldehyde, comes from ethylene. The traditional laboratory method for converting ethylene into ethanol is to react it with sulphuric acid. This forms monoethyl and diethyl sulphates which will react with water to produce ethanol and regenerate the sulphuric acid. This process was for a time used industrially but, following the trend of the modern petrochemical industry to look for the most uncomplicated method, the sulphuric acid step has been eliminated in the direct hydration process.

Water vapour and ethylene at 300 °C and under high pressure react to form ethanol. This is an equilibrium reaction – the ethanol can break down to form ethylene and water – so the conditions of the reaction are finely balanced to find the most economical conditions, which, in this case, are a 4–5 per cent conversion of ethylene to ethanol at each pass through the reactor. Certainly this yield would not satisfy the laboratory chemist, but when recycle is taken into account provides the best compromise between laws of chemistry and those of economics.

Ethylene can also be made from ethanol, and it has been suggested that, in the future, this may be a more economical source of ethylene than oil feedstock. Ethanol can be produced from a wide range of waste materials, ranging from cannery effluents to sawdust and, if large supplies of such materials could be made available on a regular basis, they might be utilized for cheap industrial ethanol production. Dr Andrew Porteous, of Britain's Open University, has devised a process for producing ethanol from domestic refuse; so it is possible that one day Britain may rely on household rubbish as a 'feedstock' for its chemical industry (which could then turn the feedstock into plastics, to make rubbish bins for storing more rubbish . . .). However, with Britain expecting to be self-sufficient in oil by the mid-1980s, as a result of offshore discoveries, that day is probably still distant.

However, in a world where people are becoming more aware of the need to rely on renewable resources, the subject is of increasing research interest. Irving Goldman, head of the Department of Wood and Paper Science at North Carolina State University, for example, told an American Chemical Society meeting in April 1975 that 50 million tons of wood – roughly the annual residual wastage of the U.S. logging industry – could provide the United States with 95 per cent of the raw material requirements for its plastics industry. According to his calculations, ethylene production via ethanol from wood will be economically comparable with petrochemical ethylene during the late 1970s. Only three months later, at a conference in London, Dr. Walter Kaminsky, of Hamburg University described research he had carried out on recovering raw materials from plastics waste by pyrolysis. At 740 °C, polyethylene, for example, yields 25 per cent ethylene, 16 per cent methane, 12 per cent benzene, and nearly 10 per cent propylene. Similar treatment of polystyrene waste leads to recovery of more than 70 per cent of styrene monomer.

The direct approach to chemicals production is by no means confined to ethanol. In at least one case its application is a threat to ethanol producers. The traditional way to make acetaldehyde in a laboratory is by the oxidation of ethanol. Some acetaldehyde has been made industrially from acetylene, but this process has been superseded by a direct one-step process from ethylene. The Wacker acetaldehyde process, which has been operational for just over a decade, consists in adding an atom of oxygen derived from a molecule of water to a molecule of ethylene. Unlike the ethanol process, the Wacker process takes place in the liquid phase and involves a palladium chloride catalyst which is converted to palladium metal during the reaction, and is then regenerated. It could only have been developed successfully in recent years, for the catalyst solutions are highly corrosive, and did not become fully satisfactory until certain parts of the plant were manufactured from titanium, a metal that was not available for such large-scale usage until recently.

The Soviet chemist Moiseyev suggested that if the Wacker process was performed in the presence of acetic acid instead of water, vinyl acetate, another polymer raw material, could be produced. This process, which is in operation on a modest scale in both the United States and England, has been intensively developed, although corrosion problems still need to be ironed out. One way to do this is to perform the reaction in the gas phase at high temperatures, rather than in the liquid phase; this modification is now being looked at very closely and could become the sole economic process for the production of vinyl acetate, since a plant could take in ethylene, convert part of it by the Wacker process to acetaldehyde (which can easily be oxidized to acetic acid), and then use ethylene and acetic acid for the manufacture of vinyl acetate.

There are two interesting stories behind this example of new chemistry in the industrial field. First, when Dr J. Smidt and his colleagues at Wacker Chemicals in Germany were experimenting with passing ethylene and oxygen over palladium catalysts, they were trying to make another raw material important for the chemical industry, ethylene oxide. When they noticed that traces of acetaldehyde were formed, their investigation revealed how the palladium catalyst was operating and, in the end, they developed the present process in which oxygen is supplied by water molecules. Second, the basic chemistry was not new at all. The formation of acetaldehyde from ethylene in aqueous solution containing palladium chloride was noticed in 1894, but had been forgotten. The real new chemistry of the Wacker process was in turning it into an economic process by working out the regeneration of the expensive palladium catalyst. This involved a basic scientific study of the mechanism of the reaction in order to find ways of influencing the reaction toward the desired products.

Not surprisingly, although the Wacker process did not produce ethylene oxide, a direct oxidation process to this compound has been developed. This involves reaction of ethylene and oxygen over a silver catalyst. Again, development of this process required a close study of the kinetics and mechanism of the reaction, for there is also a competitive reaction operating – complete oxidation of ethylene to water and carbon dioxide, neither of which are saleable commodities. On the other hand, ethylene oxide, produced at a rate of 2 million tons a year in the U.S.A., is needed for manufacture of ethylene glycol–used in automobile antifreeze and polyester fibres–and a number of important detergent chemicals, as well as for industrial separation processes.

The Wacker process can be applied to the next unsaturated hydrocarbon up the scale from ethylene, propylene. In this case it produces the solvent, acetone, another important intermediate in the production of large-scale industrial chemicals. However, the Wacker process is in competition here with a totally different procedure, the phenol-cumene process, which converts cumene into two valuable products, acetone and phenol. If it comes to a fight as to which produces the cheaper acetone, a group that will benefit in any event are the propylene producers, for propylene is also a vital raw material for the manufacture of cumene.

Although propylene is not required on such a large scale as ethylene – the 1975 production capacity in western Europe was about four million tons – it has been a fast-growing area, having quadrupled in capacity in the 1977 production figure for the United States was about 6 million tons–it has been a fast-growing area, having expanded more than 10-fold since 1950. Much of this growth was due to the requirement for polypropylene, an insignificant polymer before 1960 but now widely used, whose development involved important fundamental chemistry which won a Nobel prize for the Italian chemist Giulio Natta (see p. 155).

Another new process, propylene ammoxidation, also accounts for part of propylene's fast growth. This is now the preferred route to acrylonitrile, the basic building block for such acrylic fibres as Acrilan and Courtelle.

Like ethylene oxide, propylene is also an important raw material for a wide variety of chemicals, such as the polypropoxy ethers used in manufacturing polyurethanes. A suitable catalyst for direct oxidation of propylene to its oxide is earnestly sought, but at present the closest approach is a dual process in which another organic chemical, such as ethylbenzene, is oxidized to a hydroperoxide which reacts with propylene to form propylene oxide and l-phenylethanol. The latter can be dehydrated to styrene, required for production of polystyrene, but the success (economically) of such a process depends on the demand for styrene and the cost of alternative supplies. Other materials can be used in place of ethylbenzene (for example, isobutane, which can be converted ultimately to isobutylene or tertiary butyl alcohol), but clearly the success of the process still depends on finding a suitable market for the by-product.

One of the most successful modern processes, because of its utilization of the major by-product, must be the phenol from cumene process. Barely utilized before 1950, it now produces more than half of the world's phenol. As has been made clear, many of today's organic chemicals owe their rise to the enormous growth of the plastics industry. Phenol is no exception. About 50 per cent of the phenol made is used in phenol-based resins, while a further 20 per cent goes into the manufacture of caprolactum, a precursor of nylon 6.

The cumene-phenol process had its origins during World War II when Professor H. Hock and colleagues in Germany were carrying out a basic study of the oxidation of various organic compounds by molecular oxygen. From cumene (isopropylbenzene) they isolated a hydroperoxide which, on acid treatment, split up into phenol and acetone. To Hock's group, cumene was just one of a number of organic chemicals they were studying, and the reaction was only mentioned in one of a series of research papers on the effects of molecular oxygen. However, it was noted by industrial chemists and developed simultaneously in the U.S. and Britain into an industrial process that was first used commercially at a plant in Montreal in 1953.

Some scientists deride industry, claiming that it just makes profits from the discoveries of research workers in institutes and universities, and does not innovate itself. In the chemical industry this is untrue, for the fundamental chemical discoveries of the laboratory chemist are rarely made with any consideration for economics. Thus, the Wacker process was a piece of new industrial chemistry, for the 1894 discovery of the reaction did not include any of the work needed to overcome the cost factor of the platinum catalyst. Detailed work on the mechanism of the reaction was also carried out in industry; in 1894 the whole concept of reaction mechanisms was still uncharted territory.

Similarly, without industrial research there would have been no phenol-cumene process. Not only did the reaction involve free radicals – still a novel area of chemistry in the early 1950s – but also it had been established only that phenol and acetone were products, not that they were the sole products. Establishment of the complex series of reactions involved was carried out in industry. Accurate measurements of the number and types of products formed had to be made; the then relatively new technique of infra-red spectroscopy was used. Quantitative measurements of the rates of various steps had to be made, and it was in fact necessary to postulate reaction mechanisms that, at the time, were not known to occur. Industrialization of the process not only helped produce profits, but also added to the sum of basic chemical knowledge.

In the cumene-phenol process, a previously unmentioned raw material has been discussed – benzene. Benzene is an aromatic[1] hydrocarbon, and the aromatic hydrocarbons are also a part of today's giant petrochemical industry, in some ways complementing the smaller aliphatic molecules already mentioned. For example, p-xylene, a benzene molecule with a methyl group at each end, is the raw material for manufacturing terephthalic acid. Together with ethylene glycol – made from ethylene oxide – this goes into the polyester fibre known as Dacron or Terylene.

The aliphatic hydrocarbons discussed up to now are mainly obtained from refinery gas or natural gas, the lowest boiling fractions of cracked crude. The aromatics come in the next boiling range, roughly from room temperature to 200 °C (twice the boiling point of water). This is the naphtha fraction. Gasoline comes out of this fraction, but so do other important chemical materials.

The use of aromatic hydrocarbons in the chemical industry is long standing, especially in the manufacture of dyes. Supplies originally came from coal distillation, hence the origin of the name 'coal tar dyes' for commercial dyes in use at the end of the last century. Until recent times, coal was still a major supplier of some aromatics, notably benzene, but over the past two decades its use has declined, while that of oil has leaped ahead.

There are two reasons for the growth of oil at the expense of coal as a supplier of aromatics, both depending on technologies outside the chemical field. Coal tar, the source of coal chemicals, is a by-product of coke

1. 'Aromatic' is a historical description for those compounds based on, or containing, the benzene structure – a ring of six carbon atoms, to each of which is attached one hydrogen atom. The remainder of the carbon bonding electrons are involved in holding the ring together in delocalized orbitals.

manufacture, and the requirements for coke depend on the growth of the town gas and steelmaking industries, which have not expanded at anything like the rate of the chemical industry in recent years. Conversely, the development of jet-propelled aircraft has released a lot of petroleum aromatics that would otherwise have been required as aviation fuel.

There is, however, an enormous amount of unmined coal in the world, much of it in industrialized countries which have been hard hit by the leap in oil prices during the present decade. This has meant a resurgence of research in coal chemistry, not only to develop new routes for coal-based chemicals, but also to convert it to liquid and gaseous fuels. The South Africans have always produced their petrol and many chemicals from coal and, late in 1974, announced plans for a new £600 million plant. This will use 40,000 tons of coal a day, 28,000 for feedstock and the remainder for energy, and will probably be in production by 1980. The United States is another country investing heavily in coal (again). Between 1970 and 1975, the annual budget of the U.S. Office of Coal Research, which is interested primarily in gasification and liquefaction processes, rose from $15 million to $261 million. And it has been estimated that the amount of benzene produced from coal in the United States may increase tenfold between now and the end of the century.

A major problem faced by suppliers of petroleum aromatics was that the chemistry of the process and the demand were at variance with one another. Catalytic reforming of naphtha gives the major aromatics, benzene, toluene and mixed xylenes, in the ratio 11:55:34; demand is 58:23:19. Obviously if the demand for benzene were to be met by straightforward catalytic reforming of naphtha, there would be huge surpluses of toluene and the xylenes. To overcome this, a hydrodealkylation process was developed in which toluene and hydrogen are reacted to produce benzene.

Another coal-based aromatic which is increasingly produced from naphtha is naphthalene – the chemical component of old-fashioned mothballs. However, this is not such a growth area as the simpler aromatics, since the main product of industrial naphthalene chemistry, phthalic anhydride, can now be made from one of the xylene isomers.

Like the aliphatics, a major end-use area for the aromatics is in the plastics industry. In their various forms – synthetic rubbers, plastics, and fibres – the synthetic polymeric materials are the greatest growth phenomenon of modern chemistry.

## The plastics revolution

The first synthetic plastic was a thermosetting resin called ebonite, patented in 1843. Plastics are divided into two broad categories, thermosets and thermoplastics. The thermosets are materials which, once heated, take on a rigid form which cannot be changed by reheating, while thermoplastics repeatedly soften on heating and harden on cooling.

Much of the early work on plastics was on thermosets, and it may have been this limitation that prevented them from becoming highly acceptable materials. Another reason is that they were looked upon as substitutes for existing materials, rather than as new materials requiring new ideas in design and usage. Part of this 'substitute' outlook still survives, with people disparaging plastics as second-rate materials. However, with the major advances in formulation and use of plastics since the war, this attitude is now passing out of fashion.

Many of the early plastics were not totally synthetic, but were made from common natural polymers by chemical modification. Casein, a protein in milk, was the basis for one of them, while others were made by chemical modifications of cellulose. Celluloid, a form of cellulose treated with nitric acid to form cellulose nitrate, was widely used in early movie films. Because of its extraordinary flammability, it has been replaced in nearly all applications except the manufacture of some table tennis balls; a number of film studios have reels of old cellulose nitrate films decomposing in tubs of water to prevent their going up in flames.

The first modern synthetic plastics were discovered just before and between World Wars I and II: Bakelite, named after its discoverer, Leo Baekeland, in 1912, urea-formaldehyde resins in the 1920s and melamine-formaldehyde in 1935. Nylon, the first of the synthetic fibres, was discovered in the 1930s by Wallace Hume Carothers, who was working for the American chemical company, DuPont. Carothers set out deliberately to discover a synthetic fibre, and realized that long straight molecules would have the best chance of forming fibres if they were polymerized. Unfortunately, he died in 1937, before nylon had become a commercial article.

Carothers was undoubtedly a great chemist for, in addition to nylon, he suggested the way to make the synthetic rubber, neoprene, and also worked on a group of plastics called polycarbonates which did not reach commercial production until the 1960s, but are very useful as engineering materials because they are both transparent and extremely hard. Another range of transparent plastics discovered in the 1930s is the acrylics made from polymethyl methacrylate. Used widely in World War II for aircraft canopies, acrylics have also been used for making bathtubs, advertising signs and artificial eyes.

Undoubtedly the most successful polymer developed in the 1930s was polythene. Its discovery was an accident. There is a similarity between the discovery of polythene and the cumene-phenol process, for both turned up initially as integral parts of wide-ranging research programmes. The difference in the case of polythene was that the basic research programme was carried out in industry, and the British company where the discovery was made – ICI – followed it up.

In the late 1920s and early 1930s, very little work had been done on the effects of high pressures on chemical reactions. ICI decided to carry out a basic study of reactions at pressures above 1,000 atmospheres, to see what would happen. One of the aims put forward by Nobel laureate Sir Robert Robinson, then an adviser to the company, was to see if certain types of reaction would take place under high pressures that required catalysts at ordinary pressures. Begun in late 1931 by Dr R. O. Gibson and E. A. Fawcett, it took a year to get the apparatus and conditions right for studying these gas reactions of pairs of compounds; in November 1932 the apparatus blew up, setting the project back for several more months.

In March 1933, Gibson and Fawcett studied the high-pressure reaction between ethylene and benzaldehyde. This was expected to give either of two complex compounds. The experiment was started on Friday, 24 March. On the following Monday Gibson and Fawcett returned to the laboratory to find that there had been a leak in the apparatus and all the benzaldehyde had escaped. When they dismantled the apparatus, Fawcett pointed out that the part of the steel gas-inlet tube which had been in the reaction zone looked as if it had been dipped into paraffin wax. In his notebook Gibson wrote: 'Waxy solid found in reaction tube' – the first account of the existence of polythene. In their monthly report, they wrote: 'The reaction between ethylene and benzaldehyde has been studied at 2,000 atm and 170 °C. A waxy solid, which appears to be a high polymer of ethylene was formed.' In their July report, they wrote: 'A quantity of the waxy polymer of ethylene has been prepared and work with this reaction has now ceased.'[2] Polythene had been discovered, but why was it not exploited?

Describing the discovery in 1964, Gibson wrote:

> In retrospect it may seem strange that work should thus have stopped on what is now claimed as an important discovery, and that the

2. R. O. Gibson, *The Discovery of Polythene*, Royal Institute of Chemistry London, 1964, pp.17–18

> discovery should only have been considered of academic interest at the time. But was it so strange?
>
> At that time synthetic polymer science was only in its infancy, and the immense technical and commercial developments of recent years were not even being dreamt about. The common thermoplastics of that day were polymethyl methacrylate and polystyrene. Both these materials were hard and transparent and of very different physical properties to the relatively soft, translucent ethylene polymer we had made.[3]

Another disadvantage was that increase of pressure during the reaction led to complete decomposition to carbon and hydrogen, instead of a waxy solid. It was not until two and a half years later that the reaction was studied again, this time by another ICI worker, Michael Perrin. On 19 December 1935, he and a colleague decided to study the behaviour of ethylene at 2,000 atmospheres pressure. They added ethylene to the reaction vessel, but could not get the pressure up. They added more ethylene, although they suspected that the apparatus was leaking. When they had used up all the gas, they cooled the reaction vessel and let out what pressure there was in it. On opening it they found it to be full of white powdery polymer. It was later found that if pressure was let out before cooling, the polymer was of the same waxy solid appearance as Gibson and Fawcett's original sample.

This second round of experiments produced enough polymer to make it worthwhile for study as a material of potential commercial interest. Commerical development required further application of both science and technology. For example, Perrin found that to initiate the polymerization reaction, a small amount of oxygen – 600 parts per million – was necessary. Technological developments were required to purify ethylene on a large scale to the 99·9 per cent purity required for polymer production. The first commercial polythene plant came into operation the day that World War II was declared.

This is not the end of the polythene story, for the material produced on the basis of the ICI work, called low-density polythene, is made either by compressing the gas to 1,500–2,500 atmospheres pressure at 100–300 °C, or by polymerizing it in an aromatic hydrocarbon solution at 1,000 atmospheres pressure. In the 1950s a new type of polythene, high-density, came on the market as a result of work on catalysts called Ziegler catalysts after their pioneer, the German chemist Karl Ziegler, who shared the 1963 Nobel chemistry prize with Giulio Natta.

3. Ibid pp. 20–1.

Low-density polythene has a molecular weight of between 50,000 and 300,000, while the high-density material can have a molecular weight of up to 3,000,000. The high density material has greater rigidity and a higher softening temperature; it is made at 6–7 atmospheres pressure in a hydrocarbon solvent at 100–170 °C. In a sense, the wheel has turned full circle. Originally discovered as a by-product of research aimed at replacing catalyzed reactions by high-pressure reactions, the high-pressure process has now been matched by a catalyzed process.

However, high-density polythene has by no means replaced the low-density material. Although they are both polymers of ethylene, their characteristics are sufficiently different for them both to retain a market. The differences between them are perhaps best exemplified by their uses. Nearly 50 per cent of the low-density material is used for producing polythene film and sheet, while half the high-density material is used for blow moulding, in which air is blown into a mould containing the hot, soft plastic to produce semirigid artefacts, such as bottles. By comparison, only about 10 per cent of low-density polythene is blow-moulded. In 1978, U.S. consumption of both forms of polythene was 5.5 million tons. almost a quarter of total plastics consumption.

A close relative of polythene is polypropylene, first produced in 1954 and insignificant in the plastics market until 1960. From a figure of less than 20,000 tons in that year, U.S. production has now increased about seventy-five fold, while west Europe's consumption more than doubled between 1966 and 1970, and the 1975 figure was expected to be triple the 1970 consumption. Nearly half the polypropylene produced is used for mouldings, although fibres are also an important end-use. A recent unusual use for polypropylene has been as a synthetic seaweed. Lengths of slit film or filamentous polypropylene have been attached to the seabed around seabed oil pipelines to prevent erosion of sand from under the pipes. As they are lighter than water, the lengths of polymer float upward like ordinary seaweed. It appears that use of this material not only prevents erosion, but actually encourages the building of sandbanks.

From the end of the 1930s to the mid 1950s seems a long time between the birth of polythene and its sister, polypropylene. The apparent simplicity of extending polymerization from the two-carbon ethylene to the three-carbon propylene is deceptive; it had to await Dr Ziegler and his catalysts, and the dedicated application of these by Italian chemist Giulio Natta.

In ethylene, two carbons are joined together by an unsaturated linkage; each carbon has two hydrogen atoms attached to it. When ethylene polymerizes to polyethylene, a very long chain of carbon atoms is formed,

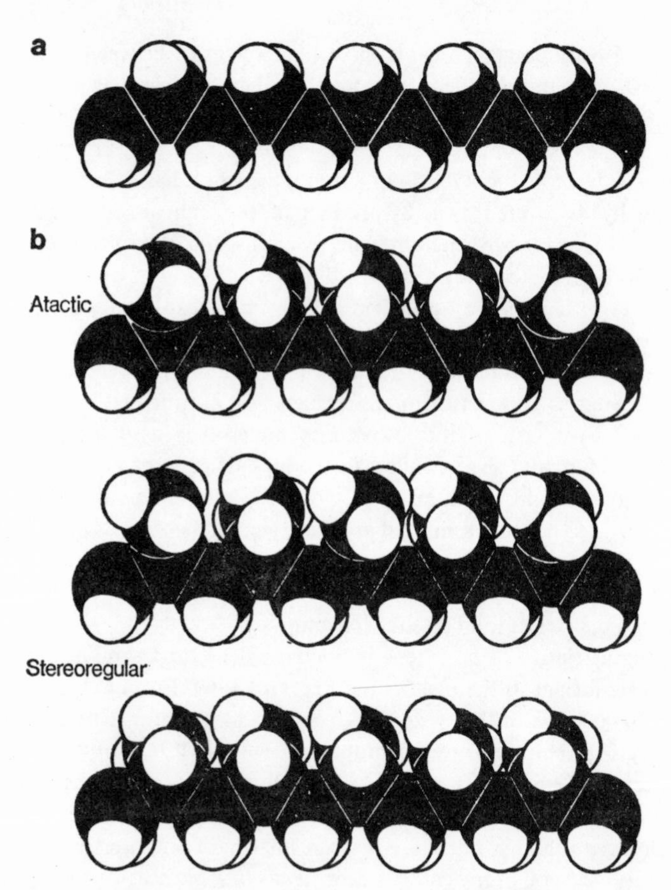


FIGURE 14 Representations of the three-dimensional structures of polyethylene (a) and various polypropylenes (b). Until the discovery of Ziegler catalysts, it was possible to produce only the disordered atactic polypropylene. Use of these catalysts by Giulio Natta led to the production of the stereoregular polypropylenes, in which the side-chains are not randomly oriented, and which have much better commercial properties than the atactic material.

still with two hydrogen atoms on each carbon atom (see Figure 14). Propylene can be looked upon as an ethylene molecule in which one of the hydrogen atoms on one carbon has been replaced with a methyl (—$CH_3$) group. When attempts were first made to polymerize propylene, the result was a polymer without useful commerical properties – it was rubbery and had little strength. Natta, a chemist who had concerned himself with determination of the structures of molecules by X-ray crystallography, showed that these properties were a result of the random way in which the propylene units joined together: there was no order about the positioning of the methyl groups sticking out of the chain. When propylene polymerizes it forms the same basic chain as polythene, but the methyl groups that replace half of the hydrogens are not part of the chain, but branch off it. Natta thought that a better polymer could be produced if the position of the methyl branches could be controlled as the polymer formed. To achieve this, he applied the work of Karl Ziegler.

As a research chemist at Marburg University before the war, Ziegler had discovered that adding metals to the free radical compounds which he was studying made it possible to form organometallic substances – molecules in which metal atoms are chemically bonded to carbon atoms. He further found that these organometallics caused some small alkenes to polymerize.

In 1943 Ziegler became head of the Max Planck Institute at Mulheim. Here he tried to polymerize ethylene using his organometallics, but it was not until 1953 that he succeeded in devising a suitable catalytic system to produce polythene. This was the basis of the processes for high-density polythene now in use, and the reason why this material has a higher density than the original polythene is because the Ziegler catalysts help to keep the growing chain straight. In the high-pressure process, the chain occasionally branches to give a two-pronged growing end. This increases the three-dimensional complexity of the molecule, and, unlike the straight chains, the chains with branches are not able to pack together tightly, thus giving a lower density material.

When a similar type of catalyst was applied by Natta to the polymerization of propylene, it produced an ordered structure in which all the methyl groups were aligned on one side of the chain. This greater regularity produced a strong polymer similar to, but in some respects better than, polythene. Low-density polythene softens in boiling water; articles made of high-density polythene or polypropylene are resistant to boiling water. Further work showed that it was also possible to produce a polypropylene in which the methyl side-chains alternated from one side to the other of the backbone carbon chain.

Ziegler catalysts have been widely used by industry to produce stereospecific polymers – those with a finely ordered structure. Nearly all natural polymeric materials are stereospecific and, with the aid of Ziegler catalysts, it has been possible to make a synthetic rubber identical in structure to natural rubber.

How do Ziegler catalysts work? Organoaluminium compounds, in which there is a carbon-aluminium bond, are reactive because this bond is not very strong. If a compound such as triethyl aluminium (an aluminium atom with three ethyl groups attached) is heated with a 1-alkene, that is, an unsaturated carbon-chain molecule in which the double bond is at one end of the chain, an insertion reaction takes place. One aluminium-carbon bond is broken, the end carbon of the alkene forms a bond with the aluminium, and the carbon fragment originally attached to the aluminium joins on to the second carbon of the unsaturated molecule. If one molecule of triethyl aluminium reacts with one molecule of ethylene, diethyl butyl aluminium is formed – one of the ethyl groups has grown into a four-carbon chain. If there are many molecules of ethylene, each of the three ethyl groups originally attached to the aluminium will undergo the insertion reaction. Once an insertion reaction is complete, it can be repeated, so that the aluminium atom will eventually have three long, straight chains attached to it. However, as the chains lengthen, they tend to break off, so the aluminium trialkyls by themselves are unsuitable as polymerization catalysts.

By using a complex of aluminium alkyl and titanium tetrachloride, stereoregular polymerization to longer chains is possible. The stereospecificity is probably a result of using the transition metal, titanium. Electrons from the alkene can coordinate to the titanium – that is, occupy its unoccupied orbitals, forming a geometrically defined complex. Held in position in this way, the alkene then inserts into the alkyl part of the catalyst complex. As its approach route has been determined by its coordination to the titanium atom, it joins in a specific way, as all previous and following alkene molecules have done or will do, thus giving a stereoregular product.

The range of synthetic polymers which are called plastics, as distinct from rubbers and fibres, is now enormous. Two other materials with a basic skeleton similar to polythene, but with different atoms branching out of the backbone, are polyvinyl chloride (pvc) and polytetrafluoroethane (Teflon). The latter, discovered in 1938, has all the hydrogen atoms replaced by fluorine atoms. Although not commercially fabricated until 1948, it was used in World War II as a coating on reaction vessels to prevent corrosion during the process for the separation of uranium-235 from the more common isotope, uranium-238, in manufacture of the atomic bomb. Its main domestic use now is in non-stick cooking utensils, where its high heat resistance and low coefficient of friction combine to ease washing-up problems.

Polyvinyl chloride was discovered much earlier – more than a century ago – but was not commercialized until the 1930s owing to the lack of a suitable plasticizer. Plasticizers are chemicals added to plastics to increase their flexibility so that they can be easily moulded or formed into the required shape. Now pvc is used for a wide range of articles – curtain rails, rigid pipes, bottles, flooring compositions, and, in recent years, clothing.

A recent development which has expanded the range of applications of plastics – notably into the construction industry – is the foam plastics. These are ordinary plastic materials which are made to foam during manufacture, so that they set into honeycombs of tiny bubbles. Polystyrene is an example. Discovered in 1839, a good commercial polystyrene was not achieved until 1937. In its solid form it has been used for the manufacture of toys, slide viewers, vacuum flasks, refrigerator components, and so on. It was not until the 1950s that expanded polystyrene was introduced, following development work by BASF in Germany. To make the expanded material, polystyrene beads or granules are impregnated with the liquid hydrocarbon pentane, then steam heated. This causes them to expand as the pentane volatilizes and escapes. The expanded material then has to be matured so that air can find its way into all the tiny crevices formed in the polymeric mass. If a moulded article is required, the expanded beads can be further heated in a mould so that they expand and fuse together to give a lightweight 'solid' object. The entrapped air makes expanded polystyrene a good insulator, and it can be applied as a sheet to walls or as tiles to ceilings, although its flammability has to be borne in mind. It is also widely used as a packaging material for glassware, cameras, microscopes and similarly fragile objects, which can be packed in boxes filled with expanded polystyrene into which shapes have been cut for the objects to fit snugly.

The other major foamed plastic is polyurethane. This is a complex polymer made by reacting two different molecular building blocks. It is possible to choose the building blocks in such a way that the reaction between them releases carbon dioxide gas, which then foams the polymer as it forms. This property has been used for *in situ* polymerization to provide insulating foams; the ingredients are poured into the cavity between an inner and an outer wall, and the polymer immediately forms and fills the gap with a foamed mass which rapidly sets solid. The advantage over the application of foamed sheet is that the plastic gets into all the nooks or crannies in the walls. Similarly, by fixing a temporary mould around pipework and filling the intervening space with polyurethane, it is possible to insulate pipework without having to dismantle it. Polyurethane can also be made in such a way that flexible foams are produced; these are used widely for seats in cars and for domestic furniture.

Perhaps the most dramatic use of an expanded polymer was in the re-floating of a sunken ship. A few years ago the *al Kuwait* sank in Kuwait harbour with a cargo of 5,000 sheep. As it sank near the inlet for the distillation plant which supplied Kuwait's fresh water, it represented a considerable health hazard. Traditional salvage methods would have taken too long to raise the wreck, so 70 tons of expanded polystyrene beads were pumped into it, and the air entrapped in these was sufficient to raise the ship. Several other sunken ships have since been brought to the surface with expanded polystyrene.

Polyurethanes can also be made into stretch fibres. The fibre industry is an important consumer of synthetic polymers. The best-known synthetic, nylon, has already been mentioned, but, in fact, there is no single 'nylon', but a whole family of them. The nylon discovered by Carothers is called nylon 6,6 because it is made from two starting materials (monomers), each of which has six carbon atoms in its molecule. One of the monomers has a carboxylic acid group at each end, the other an amino group at each end. These react together to form an amide bond of the type found in natural proteins, so that the resultant polymer is a long chain of carbon atoms with an amide linkage after every sixth carbon atom.

A number of different nylons have been tried, such as nylon 10,10 which, as its name implies, is made from two components each having ten carbons in its chain. But the most successful postwar nylon is nylon 6. This is made from a single monomer, having a six-carbon chain with a carboxylic acid group at one end and an amine group at the other. In this case, the polymer forms by monomer units linking up head to tail to form amide bonds.

Although the textile uses of nylon have been widespread – one of its major outlets during World War II was as material for parachutes – it is now used also as a solid for making plastic gear wheels and similar machine components.

In 1970 the world production of synthetic fibres was over 34·5 million tons, an increase of more than 150-fold since 1946. Immediately after the war nylon was virtually the only synthetic; by 1970 its share of the market had dropped to just under 40 per cent, the two major competitors being polyesters (about 34 per cent of the total) and acrylics (19 per cent). Synthetics, as their phenomenal growth-rate indicates, have rapidly taken over large areas of the clothing market, and it is estimated that at the present time over 30 per cent of the total world clothes fibre market are met by synthetics. As many of the developing nations will still rely almost totally on natural fibres for clothes manufacture, this means that in the developed countries a high percentage of clothing now come from oil.

Carothers's research programme, which led to the discovery of nylon, also included the synthesis of polyesters. In these the amide linkages of polyamides such as nylon are replaced by the ester linkages formed from the condensation of an acid with an alcohol. Carothers himself had no success in producing polymers that could be converted into fibres; it was not until 1941 in Britain, when J. R. Whinfield and J. T. Dickson condensed terephthalic acid and ethylene glycol (an important component of car antifreeze) to form a material now known in Britain as Terylene and in the U.S. as Dacron, that success came. Terylene does not really compete with nylon; they have sufficiently different properties that, if used to their full advantage, they appear in different types of garments. Terylene can be processed so that it has a texture more like wool than nylon does, so is better when used in suits, but as it is not so elastic as nylon, this earlier material is still used in stockings and tights.

The acrylic fibres were developed after the war; these are polymers of acrylonitrile, a compound now made mainly from propylene by the ammoxidation process which is making obsolete earlier routes from acetylene and ethylene. Nearly a million tons a year of acrylonitrile are made in the U.S., and more than 50 per cent of this goes into fibres, the remainder being used for copolymer resins such as ABS, and for nitrile rubber. Acrylonitrile polymers, known as Orlon, Courtelle and Acrilan, resemble good quality wool. A newer development is Teklan, a copolymer of acrylonitrile and vinylidene chloride which is more like silk, and which has flame-resistant properties. The acrylics are used widely for rugs and carpets as well as for clothing fibres.

Fibre uses for polyethylene and polypropylene are also being developed. Again, because of the different properties, the uses are different. Their fibre forms are found in such products as ropes, deck-chair covers, and protective clothing for workers in industry.

## Chemistry with a bit of bounce

Many of the advances in plastics, both the fibres and the bulk plastics, occurred during World War II when the shortages of natural materials required that substitutes be found. Nowhere was this more true than in the case of rubber, for most of the plantations which had supplied natural rubber fell into Japanese hands. Synthetics had been discovered before this but had not been seriously developed because, as is often the case, a new invention has to wait until it is needed before it is exploited.

Carothers's early work on synthetic fibres led to the discovery of neoprene in the early 1930s. This is the condensation product of an organic molecule called chloroprene, originally made from acetylene, but increasingly now from butadiene. Synthetic rubbers are perhaps different from other polymers in one significant respect: they are more closely allied to natural materials. Natural rubber is made, by the rubber plant, from isoprene, a molecule composed of five carbon and eight hydrogen atoms. Four of the carbon atoms are linked in a chain, while the fifth is attached as a branch to the second carbon from the end in the chain. Between the first and second, and third and fourth carbons are double bonds, which give the molecule its polymerizing capability. Chloroprene has the same basic carbon chain, but the methyl group side-chain is replaced by an atom of chlorine. If the methyl group of isoprene is replaced by a hydrogen atom, the resultant compound is butadiene, which can also be polymerized into a synthetic rubber.

Isoprene can polymerize in more than one way. Instead of forming a fully saturated chain like polyethylene, electrons from the two double bonds shift and form a double bond between the second and third carbon of the isoprene unit in the polymer chain. This means that each isoprene subunit is subject to the laws of geometrical isomerism (i.e., cis and trans forms are possible). Early attempts to polymerize isoprene led to a random polymer that was like neither natural rubber nor another natural isoprene polymer, gutta-percha. This is because both the natural products are almost stereoregular, natural rubber being a cis polymer, and gutta-percha, which is not elastic, the trans polymer. The same question of stereospecificity applied to the polymerization of butadiene; it was not until the development of Ziegler catalysts that a useful butadiene rubber could be produced.

After the war, during which commercial uses and production of synthetic rubbers were developed, it became clear that natural rubber would never again be able to meet the world demand for rubber. As a further development of the uses of Ziegler catalysts, it is now possible to manufacture 'natural' rubber by the controlled polymerization of isoprene. This has been done on only a small scale so far, despite the fact that the rubber is a little superior to the truly natural material. The Ziegler catalysts are even more stereospecific than the natural catalysts in the rubber plant, and produce a polymer more stereoregular than the approximately 90 per cent regular plant rubber.

The greatest interest in synthetic rubbers has not been to find ways to beat nature at her own synthetic work, but to produce rubbers that are better suited than the natural material to man's needs. By 1976 the non-Communist world production of rubber amounted to about 8 million tonnes, of which 67 per cent was synthetic; by 1986 synthetics should account for 69 per cent of the market, which is then expected to be in excess of 13 million tonnes.

The major user in the U.S. is the car industry, tyres and tyre products consuming more than 60 per cent of the rubber used. One of the major rubbers in the tyre field is butyl rubber, a copolymer of isobutylene with a small percentage of isoprene needed for 'vulcanizing' the rubber. Ordinary natural rubber is not much use as an article of commerce because it tends to lose its elasticity in the cold and go sticky when hot. This problem was overcome in the 1830s by Charles Goodyear, who found that a much less changeable material could be produced by heating rubber with sulphur, a process he called vulcanization. The sulphur reacts with the double bonds in the polymer chain to form bridges between parts of the chain (somewhat analogous to disulphide bridges in proteins). Isobutylene has only one double bond, so that when the molecule polymerizes, it is fully saturated and is not susceptible to vulcanization unless a small amount of isoprene is added to keep some double bonds in the final polymer.

The original market for butyl rubber, tyre inner tubes, disappeared with the invention of the tubeless tyre; however, it is now used extensively as liner material for tubeless tyres, where its low air permeability and resistance to oxidation make it particularly suitable. The major component of tyres is styrene-butadiene rubber, a 1:3 copolymer of styrene and butadiene. About two-thirds of U.S. consumption, well in excess of a million tons a year, goes into tyres. It is also used for foam rubbers and in shoes, although its growth potential has been sapped by development of the stereospecific rubbers.

One of the toughest rubbers around is nitrile, made from acrylonitrile and butadiene. This became important during the last war for manufacturing self-sealing petrol tanks. It is very resistant to organic fluids and is now used for applications such as petrol pump hosepipes.

The plastics story does not end here; in a sense it is now unending, with new plastics coming up every few years. The thermosetting plastics, the original of which was Bakelite, have not been mentioned. Most of these were discovered in the 1920s and 1930s and have not shown the great rise experienced by the thermoplastics, fibres and rubbers. They are, none the less, an important part of modern life, and the raw materials are petroleum based. The silicones, a range of plastics in which alternating silicon and oxygen atoms rather than carbon form the backbone of the polymer chain, are new, but they have not been discussed in this chapter because they are not part of the petrochemical industry.

This chapter has tried to follow a thread which joins the basic laws of chemistry to those of the marketplace. The new chemistry of industry may be old chemistry to the academic researchers. But even if he uses old chemical knowledge, the unique criteria which he must apply make the industrial chemist change this chemistry to suit the needs of his company. One of the keynotes of chemical industry, which has been obscured by the approach adopted here, is its ubiquity. Chemical products find themselves everywhere, pressed into service not only in the home but in dozens of industries. Some of the ubiquity of modern industrial chemistry is described in the next chapter.

# *CHAPTER 6*

## CHEMISTRY IS UBIQUITY

Wherever we look we see the effects of modern chemistry. This is not surprising for, in so far as every material thing is made of molecules, it is 'chemical', and all modern artefacts are, in a sense, the product of chemistry. However, it is misleading and unjust to other branches of science to claim many modern inventions as triumphs of chemistry.

Take, for example, the transistor, which may appear in future histories as the most significant single invention of the twentieth century. The first transistor was demonstrated on 30 June 1948, by three scientists from Bell Telephone Laboratories: William Shockley, John Bardeen and Walter Brattain. For their discovery they won the 1956 Nobel prize for physics. This was appropriate, for the transistor as an invention sprang from physics rather than chemistry, although what has been described as 'the breakthrough that was to take transistors out of the laboratory and put them in every home',[1] zone refining, discovered in 1952 by William G. Pfann, is often considered a chemical separation process, and is probably more widely used in chemical laboratories than anywhere else.

Similarly, advances on the transistor principle, which have led to integrated circuits and microminiaturization, without which the present and predicted future ranges of computers would be impossible, are all basically physical. The contribution of chemistry has been to select the materials which fit the theoretical background worked out by the physicists, and for physicists, chemists, and engineers working in collaboration to develop this new technology.

There are other areas in which it is clearly the contribution of the chemist that has led to modern advances. Four out of every five drugs now in use were unknown thirty years ago, and many of the most successful modern drugs – and other medical aids, such as surgical spare parts – are the result of chemical research. Here the chemistry always has to be coupled with co-operation from doctors and, in the case of drugs that have been isolated initially from microorganisms, with microbiologists. The future progress of man depends on breaking down the old subject divisions

1. David Fishlock, *The New Materials,* Basic Books, New York, p. 148.

of science and the creation of interdisciplinary research and development. Consequently, it is difficult to provide a coherent picture of the limits of chemistry in modern life. As we move away from the clearly defined centre, exemplified by the petrochemicals operations described in Chapter 5, distinctions between chemistry and other sciences blur to an even greater extent.

Take the field of materials science. Until the 1930s there had been very few new materials developed by science for mankind's use. Building bricks were still made on the same principles used by the ancient Egyptians, clothing came almost wholly from natural fibres processed in ways that showed clear kinship with prehistoric methods. Various changes in economic and social structures, such as the war-induced shortage of natural rubber, led to the development of embryonic new polymers. Development of the jet engine led to a need for new, high-temperature alloys.

Since the end of the war, materials science as a subject in its own right has grown phenomenally. At first, new materials such as plastics were misapplied as substitutes for natural materials – for example, plastic imitations of bone handles for knives. The true development of materials science came when engineers, physicists and chemists merged their talents to produce synthetic materials for specific purposes which could not be met satisfactorily by natural materials. This 'molecular engineering' has led to a whole new range of compounds and of composites, specially designed mixtures of compounds which have properties that excel those of either component on its own.

The hardest natural material known to man is diamond, a polymer of carbon atoms in which each atom is bonded to four other carbon atoms to form a regular matrix based on the tetrahedron. Other forms of carbon also occur naturally, but these have different structures; graphite, for example, consists of flat sheets of interlinked carbon atoms which can slide over one another, thus giving the material a characteristic slippery feel. In 1955, General Electric scientists found that graphite could be converted into diamond by heating it to a very high temperature under a pressure 100,000 times greater than normal atmospheric pressure. Diamond, at that time, was greatly in demand in industry for grinding and cutting operations, and a synthetic diamond that could be manufactured at a substantially lower price than mined diamonds was a profitable prospect.

Once the molecular engineers looked at the characteristic property of diamond and related it to its tightly knit molecular structure, it became clear that it should be possible to make other very hard compounds. Two years after its diamond synthesis, GE produced borazon, a polymer made from equal parts of boron and nitrogen. These two elements are placed on either side of carbon in the periodic table; one has one electron less, one an electron more, so that the overall result of polymerizing a 1:1 mixture is to get the same electronic configuration as carbon polymer. The GE scientists started their experiments with boron nitride that was like graphite; when they finished, they had produced boron nitride like diamond. This material was not only like diamond in its hardness, in one respect it was better. At high temperatures carbon reacts with oxygen in the air to form the gas carbon dioxide; borazon is less reactive and therefore better for high-temperature grinding operations.

This early demonstration of molecular engineering seems unsophisticated by comparison with some modern examples. In 1967 GE developed a new technique called metalliding for diffusing boron atoms, with the aid of an electric current, on to a molybdenum surface. The effect of this was to make the normally soft molybdenum nearly as hard as diamond by forming a surface layer of tough alloy.

Other techniques for hardening materials have also been developed; most spectacular are the composite materials which may replace metals. In manufacturing composites, very strong 'whiskers' of material such as silicon nitride or carbon fibre are embedded into resilient resinous material. The whiskers or fibres provide strength, while the matrix in which they are bound provides the cushioning needed to prevent these strong but brittle materials from fracturing. A recently developed GE composite – using silicon carbide and silicon – combines many of the properties of metals with temperature and corrosion resistance which means that it could be used at temperatures 200 °C higher than the limit for metals and alloys.

Although some composites – glass-fibre reinforced plastics, for example – have been in use for some years in building small boat hulls or car bodies, the real advances that will come from the use of composites have not yet been achieved. Recent developments indicate that the technique may be extended to traditional materials; work is in progress on carbon fibre reinforcement of metals and glass. However, there is still more to learn before the exact advantages and limitations of these materials are known. An example of the dangers of exploiting new technology too soon was the inclusion of a carbon-fibre strengthened material in the specification for the fan blades in the Rolls-Royce RB211 jet engine. This engine, designed for the Lockheed 1011 Tristar airbus, had to be modified when it was found that accidental entry of an object, such as a bird, into the engine when travelling at high speed could cause shearing of the fan blades. As a result, Rolls-Royce had to fall back on a metal, titanium, at much greater cost. The subsequent bankrupting of Rolls-Royce in 1971 and the repercussions on the American aero-industry, which stem at least in part from this too imaginative leap into the future, made headlines throughout the world.

The 'whiskers' which have been developed in recent years are specially prepared samples of nitrides, carbides and oxides – direct chemical combinations of nitrogen, oxygen, or carbon with another element. Oxides, embedded in a matrix, have been used for centuries to confer strength on materials, although it is only recently that it has been realized that this is the case. When the oxide of silicon, silica, is heated, it forms a glass. Although glass appears solid, it is actually a supercooled liquid. Careful measurements have shown that glass flows, albeit slowly. In some European cathedrals, individual panes in stained glass windows are thinner at the top than at the bottom because of this slow flow. Glass is also brittle; ceramics, such as porcelain, are less brittle because they are reinforced by small particles of other oxides.

Because they generally withstand high temperature, much research has gone into the design of ceramics. One which has had widespread application in the home is glass ceramic, which was discovered as the result of an accident with some photoform glass. This contains metal ions which, on exposure to ultra-violet light, cause devitrification of the glass and make it opaque. As devitrified glass is less resistant to acid attack than ordinary glass, the photoform material can be selectively exposed to radiation and the exposed parts eaten away with acid; hence the name photoform. When a sample of photoform glass was inadvertently overheated, it was found to have developed strength more appropriate to ceramics than glass, due to the formation of small crystal nuclei in the glass matrix. Such glass, marketed as Pyroceram (U.S.) and Pyrosil (U.K.), is very useful for the fabrication of ovenware which will not crack when transferred from hot to cold surfaces.

Development of glass ceramics is not the only postwar innovation in glass technology. Although glass has been known for thousands of years, it is only in recent decades that a scientific understanding of its nature has been obtained. The result of this has been the extension of molecular engineering to this old material, leading to controlled changes in glass which give it new uses.

The whole technology of glass manufacture has undergone a revolution in the past two decades. Most of the glass manufactured is plate or sheet glass. In order to obtain flat glass in which both sides of the sheet are parallel, it used to be necessary to grind and fire-polish the rough sheets of glass produced from tanks of molten material. At the end of the 1950s an English company, Pilkington Brothers, introduced the float glass process which is now used throughout the world. Molten glass is passed from a furnace on to a bath of molten tin. As it passes across the flat tin surface, the glass forms a sheet of even thickness which gradually solidifies into perfect plate glass.

One of the advantages of this process is that it can be used for making special glasses with unusual properties. Different coloured glasses, for example, are made by addition of metals to glass. In the Pilkington process an electric current is passed through the molten glass as it floats on the tin; this current can be used to draw metal ions through the glass until a desired concentration of a particular metal is reached. Because the Pilkington process operates in an atmosphere of hydrogen (to prevent oxidation of the molten tin), the metal ions are converted to free metal atoms – the form which imparts colour – in the glass.

In the early 1960s a new type of glass which is reactive to light was formulated. This photochromic glass contains chemicals which react reversibly with light. When the glass is exposed to strong light it darkens, but in the dark it turns transparent again. Photochromic glass can be used for prescription spectacles that enable the wearer to do away with 'clip-on' sunglasses, aircraft windshields, or computer graphic display systems, but the most important potential use is probably in data storage devices. Photochromic material can be 'exposed' like a photographic plate, and microimages stored in this way (ten copies of this book could be fitted easily on to a 10 cm × 15 cm plate), or erased if no longer required. The secret of photochromism lies in the light-reactive molecules which are blended into the glass or, more likely in the case of data storage, transparent plastics. Molecular structural changes such as isomerism and ionization, which result in a colour/no-colour transition, can be produced in particular molecules by radiation of specific wavelengths, as happens in the case of the visual pigment, rhodopsin.

One of the problems of photochromic glass has been slow response time. The chemical reaction involved has usually been dissociation of a silver compound – the darkening of the lens being due to formation of silver particles. While this happens quickly, recombination is a slower process, so that 'clearing' of spectacles on going indoors after being in bright sunshine can be irritatingly slow. A system with a much faster response time was launched in April 1975, in a bizarre application: anti-nuclear visors, to protect military personnel from the flash of a nuclear explosion. The photochromic agent in these is an organic chemical, developed in Sir George Porter's laboratory at the Royal Institution in London. It darkens in a millionth of a second, and clears in about half a second.

## More on polymers

The new materials which really dominate our society are the organic polymers. In Chapter 5 some idea of the range of these – fibres, hard and soft plastics, and synthetic rubbers – was given. Annual world production of all these materials is approaching a hundred million tonnes, and could be in excess of 400 Mt/a by the end of the century. It is probable that most of this growth will be in larger-scale production of known polymers rather than in newly invented ones. Polymer manufacture on a large scale now requires such large investments that the economic odds against developing a new polymer for tonnage applications are astronomical. Consequently, research is devoted largely to discovering how to modify existing polymers to extend their range of uses, or to the development of high value polymers for small-scale, specialist applications. Two examples of the latter field are the development of plastics for use at high temperatures, and electrically conducting plastics. In the autumn of 1964 John Lupinski and Kenneth Kopple first described a truly conducting plastic. Research is now focussed on development of a semi-conducting plastic material which will compete with such materials as silicon and gallium arsenide in the manufacture of advanced electronic components.

With high-temperature plastics, the molecular engineers are further ahead. For more than a decade now we have had 'ladder polymers' in which two-stranded polymer chains are held together by cross-links, like the rungs of a ladder. The instability of ordinary polymers relative to heat is caused by the breaking of bonds in the polymer chain, which leads to much shorter molecules. In a ladder polymer, when a chain breaks in two, the rungs holding it to its sister chain still keep the parts together.

A major success story of the polymers – if one can pick out a single class – is the silicones, polymers whose backbone is based not on carbon atoms, but on alternating atoms of silicon and oxygen. If this were all there were to the molecules they would be inorganic, but silicones also have organic side-chains. Like so much of modern chemistry, the division into inorganic and organic branches has little meaning here.

In 1943 Dow Chemicals and Corning Glass formed the Dow Corning company to exploit silicones; in the same year General Electric began to market a silicone rubber. This was only two years after Eugene Rochow – now at Harvard University – had discovered the basic process for manufacturing methyl silicones, that is, silicone polymers in which the side-chains are methyl ($CH_3$) groups.

Although silicon occurs directly below carbon in the periodic table, and is therefore in the same 'family', it does not have carbon's power of combining with itself. Bonds between silicon and other silicon atoms are easily broken, as are bonds with other elements such as nitrogen and sulphur. In addition, silicon bonds are too long to allow the orbital overlap necessary for double bonding. This means that the bonds characteristic of carbon-based living molecules are unable to form. However, a repeating chain of silicon and oxygen atoms is stable, and silicon is like its relative carbon in generally forming four bonds – so there are two side-groups attached to each silicon atom in a silicone, as the polymers are called.

By varying the chemical nature of the side-chains, it has been possible to produce a range of silicone fluids for uses in hydraulic systems, skin lotions, toothpaste, car and furniture polishes, and as antifoams for large-scale pharmaceutical fermentations. Silicone greases, combinations of liquid silicones with solid silica, are electrical insulators, as are the silicone rubbers formed by cross-linking silicone chains in which some of the organic side-groups are attached to two silicon atoms on different chains (this cross-linking is similar to that produced by the vulcanization of natural rubber).

It is estimated that a modern large aircraft contains about half a ton of silicone rubber. Because of its non-carbon backbone, silicone rubber is non-flammable; the U.S. and British navies use it as the covering for all electrical wiring in ships. In the event of a fire, the silicone is converted into a non-conducting coat of silica.

Cross-linked silicones are effective water repellents, and are used for treating masonry, textiles and paper. An advance on this technique is silylation, for making glass, masonry and minerals water repellent. These materials contain reactive hydrogen atoms which are replaced by trimethylsilyl groups, thus giving a surface coating of strongly bound, water-repellent, organic material.

Possibly the most successful applications of silicones have been in medicine. Because all living organisms have evolved active defensive systems against foreign materials, it is difficult to manufacture suitable spare parts: most organic plastics cause unpleasant reactions if inserted in the body. Because silicones are so 'unnatural', the body does not recognize them, and consequently does not usually react against them. This means that it is possible to implant silicone materials without the danger of unpleasant side-effects.

Silicones have been used for replacing defective heart valves and for enlarging the female breast, but the most important medical application is probably the Holter ventriculo-caval shunt. This is used for draining excess water from the brains of babies suffering from hydrocephalus. This condition, which consists of a build-up of water pressure in the brain, causes babies to develop overly large, distorted skulls. The Holter shunt prevents this by providing a drain for the excess water. It passes through brain, muscle, skin and bone, and parts of it are permanently bathed in blood, yet its undoubted success is testified to by the fact that many thousands of these shunts have been inserted since their invention.

## Getting enough energy

A major theme in this book has been interactions between matter and energy. These two cornerstones come together once more when we look at the technology of modern life, for energy supply is a fundamental aspect of civilization. Without regular supplies of energy in diverse forms, many of the new materials at our disposal would be of little use, for they have been designed to fit in with our energy utilization patterns.

Until recently, most of man's energy supply came from burning fossil fuels – coal, oil and natural gas – and using these to generate heat, electricity, or, as in the automobile, motion. For decades warning signs have been up that this energy supply will run out – or at least run so short as to become rapidly more expensive.

This has led, in the field of power supply, to the development of nuclear reactors which use the energy of decomposition of radioactive elements for generating electricity. Nuclear engineering, like so much of modern technology, is a multidisciplinary exercise calling on the talents of chemists, physicists, metallurgists and engineers. In outline, a nuclear reactor contains fuel elements – often ceramic materials, carbides, or oxides of radioactive elements such as uranium and plutonium. When the radioactive atoms break down, the energy released is used to heat a liquid such as water, and the heat used to generate electricity. Water, for example, is converted into steam, which drives turbines.

This is not the only way in which to harness nuclear energy. For example, the heat from the nuclear processes may be used without further conversion, perhaps as a power source in flash distillation plants for water purification.

Work is currently progressing on nuclear reactors which convert uranium-based fuels into plutonium, which can itself be used as fuel. As more usable fuel is produced than is put in, these are called breeder reactors. France's Phénix reactor started up (went critical) in August 1973 and reached its design output of 251 MWe on 12 March 1974, nine days after Britain's Prototype Fast Reactor at Dounreay went critical.

Although Dounreay's reactor has since been shut down, commercial fast breeder reactors may be in operation within a few years.

Civil nuclear energy began to develop rapidly in the mid 1960s, almost a decade after the first civil nuclear electricity generating plant had been opened in the Soviet Union. It was not until 1958, four years after the Soviets and two years after the British, that the U.S. got its first nuclear power plant for civilian electricity production at Shippingport, Pennsylvania. However, a year earlier, the U.S. Navy launched the world's first nuclear powered ship, the *Nautilus.*

In 1968 it was predicted that, by the end of the century, half the electricity generated in the U.S. would come from nuclear sources. Since that time, however, there has been rapidly growing concern, not restricted to the U.S., about the safety and environmental aspects of nuclear power, and this 'environmental lobby' may slow down – or even halt – the growth of this form of energy generation (see Chapter 7).

The demand for more energy is one of the gravest problems facing mankind, yet, in purely scientific terms, we waste far more energy than we use. When a power plant burns fossil fuels, the chemical energy locked up in the fuel is converted to thermal energy; this thermal energy is converted to mechanical energy, which is used to generate electricity. The efficiency of this three-step process is low, and when, in the end, the consumer uses the electricity, he generally converts it back to heat or motion, again at less than 100 per cent utilization, so that yet more of the precious resource is wasted.

One way in which the efficiency could be increased is to cut out the middle man. This is the basis of the fuel cell, which converts chemical energy directly to electrical energy, and may become an important energy source in developed countries in the next few years. The first fuel cell was described in 1839 by Sir William Grove, a British chemist, but it was not until 120 years later that another British scientist, Francis Bacon, developed the first modern fuel cell. This modest cell generated 5 kw from the reaction between hydrogen and oxygen.

Basically, the fuel cell generates electricity in the same way as the ordinary battery. The major differences are that fuel is continuously supplied to the electrodes, and that the power produced by the cell is a function of its size. Because they can work at very high efficiencies, fuel cells have played an important part in supplying electricity to space capsules in the U.S. space programme.

In 1961 K. R. Williams and P. D. Gregory patented a microporous plastic electrode matrix on which metal and catalyst could be deposited. This enabled high currents to be drawn from fuel cells, and also permitted the use of ordinary air rather than oxygen in oxidation reactions. The second major breakthrough was the development of systems that did not require hydrogen gas, but could use alternative fuels as sources of hydrogen. Methanol is one such fuel. The methanol/air fuel cell produces carbon dioxide and water as waste products, just as combustion of methanol would; however, the efficiency of electricity generation is much higher. Hydrazine is another possible source of hydrogen ions. Hydrazine fuel cells give off nitrogen and water. Unfortunately, this substance is not only expensive, but also explosive and poisonous.

The major drawback in fuel cell utilization at the moment is that many systems have to run at high temperatures, thus making the present generation of fuel cells unsuitable for such uses as powering cars. However, there seems little doubt that, as technology improves, fuel cells will be used in transportation and possibly in integrated energy systems for the home – where generation of all the energy he needs is left to the consumer and his private power plant.

If enough power could be generated, direct conversion of chemical into mechanical energy might also have a bright future. Professor Aaron Katchalsky, who was a victim at the Lod Airport massacre in May 1972, worked on chemical engines for a number of years. In 1950, together with Werner Kuhn, he demonstrated that mechanical work could be obtained from polyelectrolytic compounds such as collagen – the protein found widely in man and animals as connective tissue. Polyelectrolyte fibres can be made to stretch or contract according to their environment. For example, a 10 cm strip of collagen can lift 7,000 times its own weight through a distance of 3 cm when treated with the right chemical solution. If lithium bromide is used as the contracting agent, contraction takes only 0·1 seconds.

At his laboratory in Israel, Katchalsky produced chemical engines in which a belt of polyelectrolyte passes through a bath of concentrated salt solution and, later, through water. The result is to make a section of the polyelectrolyte – that in the salt solution – contract; but when a contracted portion of the belt passes through the water, so that the salt is diluted, it expands. This alternate expansion and contraction of sections of the belt can be used to make it turn around continuously, using up the energy that is normally dissipated when a strong chemical solution is diluted.

Chemistry may also play an important role in the development of other forms of energy. If it were possible to trap and store energy from the Sun effectively, mankind's dependence on fossil fuels (which trapped and stored that energy millions of years ago) could be lessened greatly. The best way to do this may be to develop photochemical systems, broadly analogous to those used in plant photosynthesis, and store solar energy in molecular packets. Another possibility, of course, is to pay more attention to the products of photochemistry which we currently waste. About five million tons of straw are burnt in Britain each year, representing an energy loss 25 per cent higher than the petrochemical-based energy input to the whole of British agriculture. It has been estimated that half the energy value of this straw could be captured by fermenting it to produce liquid fuels. Energy could also be obtained from other forms of waste. The possibility of producing petrochemical feedstocks from domestic rubbish was mentioned in the preceding chapter (p. 146). It has also been calculated that Britian's 20 Mt/a of rubbish could be burned to provide the same amount of energy as 6 Mt of coal. However, before adopting any schemes to obtain energy from such unusual sources, it is essential to draw up a proper energy account. For example, how much extra energy would be required to deliver the rubbish to the small number of large-scale incinerators that would be needed to handle it, compared to the amount currently used to transport it relatively short distances to municipal dumps?

## A drop of 'oil'

For the moment, the possibility of cars and other engines being driven directly by chemical energy is probably far-fetched. But there is another way in which modern chemistry helps to overcome inefficiency in energy-transfer processes. Wherever mechanical energy is used, part of it is wasted as thermal energy because of the friction between moving parts. To keep turning effectively, the wheels of home and industry need lubrication.

When any two surfaces move past one another, there is friction – resistance to the movement. The resistance depends partly on the materials involved and on their surface properties. For example, your hand slides more easily across a piece of polished or painted wood than across a brick wall. When two surfaces are in contact and in motion for much of the time, lubrication is necessary. In a motor, where metal parts are moving past one another all the time, a thin film of oil prevents direct contact and stops the machine seizing up.

But why should a machine seize up? The basic reason is that two smooth pieces of metal, such as ball bearings, are in reality very rough on a microscopic scale, with surfaces made up of myriad minute hills and valleys. When two such bearings come into contact, the only points that touch are a few of the hills on each bearing. If the bearings are under load, although this load may be small for the bearing as a whole, channelling it all into those few hills puts each under immense pressure which may cause them to weld together. If only small welds are formed, the energy of motion will overcome them, breaking the weld; but this exposes a fresh piece of metal, changing the contours once more.

Since the advent of the large engine during the nineteenth century, mineral oil lubricants have been used to keep wheels turning. In recent years a new concept of molecular lubrication has come into fashion. While mineral oils act by producing a thin film which coats metal parts, the new molecular lubricants react with the metals themselves, so that the surface of the metals becomes lubricating.

General Electric, the largest manufacturer of rotating machinery in the world, established in 1958 a special lubrication research group to look into the whole question of lubrication. Headed by Dr Arthur M. Bueche, now Vice-President of the GE Research and Technology Center, this interdisciplinary group set out to look at the fundamental sciences underlying lubrication. One area they chose to examine was the behaviour of freshly exposed metal surfaces.

Most of the metal we see around us, bright and shiny though it may appear, has a surface chemically different from its middle. Many metals react with oxygen from the air to form metal oxides, but once a coat only a few hundred thousandths of an inch thick has formed on the surface of a metal, the remainder is protected. Some metals also adsorb gases strongly, forming loosely bonded complexes with them. It was this latter property that opened one new approach to lubrication.

Using ultra-high vacuum to prepare clean metal surfaces and keep them uncontaminated, the GE team was able to show that aluminium has a strong affinity for alkenes. When alkenes were chemisorbed on fresh aluminium, they clung tenaciously. Not even the temperature of boiling water could remove them. Since, during the friction process, clean surfaces are continually produced as miniature welds break, the GE scientists decided to compare two mineral oil-like long-chain hydrocarbon molecules as lubricants for aluminium. The only difference between the two was that one had a double bond in it. It was easily the superior lubricant, presumably because it was instantly adsorbed on to freshly exposed metal surfaces.

Not all the metal in a bearing is freshly exposed; part of it will already be oxidized. During the friction process, tiny particles of metal oxide can break off to form a hard grit in the lubricant which provides cutting edges that damage the bearings. By altering the molecular structure of the alkene lubricant so that it was also attracted to oxide particles, it was possible to improve its lubricating properties even more.

Many metals are very reactive when a fresh metallic surface is exposed, and some will react very quickly with iodine to form solid iodides which have a lamellar structure; such structures slide over one another readily. Consequently, it seemed likely that iodine would be a useful additive to conventional mineral oil lubricants, because it would mop up exposed metal surfaces and make them self-lubricating. Iodine does not dissolve in most mineral oils, but it readily forms molecular complexes with organic compounds like benzene. Consequently, such complexes – which will dissolve in mineral oil – were added to conventional lubricants. Once again the theory proved correct, and another new idea in lubrication was born.

The whole science of friction, lubrication, and wear – which has been given the name tribology – has undergone a revolution in the past few decades. Lubrication is not restricted to preventing seizure and unnecessary wear of moving parts. There are cases where it is used to promote wear. In a new car the freshly machined parts do not 'mate' well because of the difficulties of mechanical machining. By using a highly reactive extreme-pressure lubricant, it is possible to improve the mating quickly. Lubricants of this type are organic compounds containing sulphur or chlorine. Although stable at room temperature, the molecules break up when heated to give off sulphur or chlorine. As high spots on moving surfaces are subject to the most friction, these get hotter than the remainder of the metal. The lubricant releases sulphur or chlorine on to these hot spots, which then react to form sulphides or chlorides. These are softer than the metal, so they break down more rapidly, thus smoothing out the roughness in the machined parts.

An important aspect of lubrication is that it should be effective over the whole range of operating conditions of the equipment it lubricates. An ordinary mineral oil, when warmed, rapidly becomes much less viscous; in a car it will be too stiff for satisfactory performance when the engine is cold, and too runny when the engine is hot. Consequently, modern lubricants are a complex mixture of chemical substances in a mineral oil base. In the 1950s multigrade oils were introduced, the viscosity of which changes only slightly over a large temperature range. The 'magic ingredient' is a polymer, usually a polymethacrylate or polyisobutylene. When the oil is cold, the polymer molecules form random coils. As the oil heats up and its viscosity decreases, the polymers stretch out; this has a viscosity-increasing effect and thus counteracts the thinning of the oil.

### Thought for food

All of the materials and energy sources discussed so far reflect man's basic drive for pleasant shelter from a hostile environment. But his most basic need is to eat and drink. If he chooses carefully where he is to live, he may be able to do without clothing and shelter. Without food and drink, he will rapidly cease to exist. Because this addiction to eating and drinking has been around for a long time, much of the craving has been attended to by methods that involve no obvious chemistry. However, there are ways in which chemistry helps to keep us well fed, and there is a significant difference between just being fed and being well fed.

At the moment, many of us are ill fed. This is true in the developed no less than the underdeveloped countries. While those in underdeveloped countries starve for lack of food or develop deficiency diseases because their diets lack some essential ingredient, thousands of people in the developed countries shorten their lives by overeating. Chemistry can help at both ends of this spectrum.

In increasing food output, the chemical industry has been helping the farmer for years with the supply of fertilizers. As plants contain their own mechanisms for producing the complex organic molecules which they need, fertilizers are generally mixtures of inorganic chemicals which supply the required amounts of elements such as nitrogen, phosphorus and potassium to soils. Although this aspect of chemistry goes back to the turn of the present century, there has been continued growth in soil science. Farmers can now obtain fertilizers specially formulated to suit the soil on individual farms. In addition to nitrogen (N), phosphorus (P) and potassium (K), these contain small quantities of elements that are lacking in the soil but are essential, albeit in tiny amounts, to the optimum growth of crops. World production of fertilizer is on an enormous scale: in 1972–3, production of N, P and K were, respectively, 36 million tonnes, 25 million tonnes, and 20 million tonnes. North America and Europe produced more than three-quarters of this total.

This aspect of modern chemistry is a logical growth of the agricultural sciences that originated more than a century ago. Aspects in which new chemistry is involved are the 'chemical ploughing' of land to prepare it for seeding, and the preservation of crops from insect pests. The growth of pesticide usage has been rapid and, in some ways, disastrous – a subject taken up in Chapter 7. Although it is just over three decades since the first of the synthetic pesticides, DDT, was developed, it is already banned in many places because of environmental hazards. It seems that many of the present pesticide formulations will give way in the next few years to more cleverly designed systems, such as those mentioned near the end of Chapter 4.

One of the prime purposes of ploughing land is to free it from weeds. In the last few years it has been possible to kill all plant life in a field and within days plant seed that will grow into a healthy crop. This is done with chemicals such as paraquat and diquat, simple organic molecules which interfere with the photosynthetic mechanisms of plants. The chemicals can be sprayed on to a field, where plants take them up through absorption and, depending on the weather (which affects the rate of photosynthesis), show signs of death in one to three days. The great advantage of these chemicals is their molecular structure; they are flat and electrically charged, which means that they adhere strongly to clay particles in soil, which also have flat surfaces and electrical charges. Because of this, the molecules of herbicide are inactivated and cannot harm any plants which grow subsequent to their usage. As ingestion of paraquat is nearly always fatal, this inactivation mechanism is of prime importance to humans.

Another advantage of this system is that it leaves the surface of the soil undisturbed in a way which ploughing does not. Because of the higher content of organic residues from dead plants, the surface of a soil may be less susceptible to erosion than material from below the surface which is exposed by the plough. So 'no-tillage' systems of husbandry can conserve soil as well as making land in such areas as steep slopes, which cannot presently be ploughed, suitable for cultivation.

These herbicides are not the whole answer. As the demand for food increases world-wide, land is being brought under cultivation which in the past was considered unsatisfactory for farming. There are many reasons why land may not be ideal for growing crops: mineral deficiencies in the soil, poor soil structure and bad drainage are among them. Chemists are looking at all of these in an attempt to develop conditioners which will improve or stabilize soils. Both polymers and rubbers have been used to stabilize soils, in some cases too successfully; bitumen emulsions were found initially to make the soil waterproof. This problem has now been overcome and these materials are being used to reclaim large areas of unstable land by establishing grass on it.

One answer to these very complex problems is to do away with soils altogether. In a sense this is two answers, for doing away with soil can mean growing conventional plant foods without soil, or producing unconventional foods. The former (hydroponics) has been tried, especially in countries with large arid regions of poor soil, such as Israel. Basically, the plants are fed a carefully controlled composition of chemical solution while resting in a synthetic soil substrate, or in no substrate at all. Strange though the latter may sound, plants which grow upright can be supported in such a way that the roots hang free, gathering nutrition from pipes of chemical solution.

The alternative is to produce new foods – edible materials which have not been used on a large scale by man before, such as yeasts, algae and bacteria. Micro-organisms have, of course, played an important role in human diet for thousands of years – for they are essential to the production of a wide range of foodstuffs, such as bread, cheese and yoghurt. However, large scale production of these materials, using non-traditional starting materials, is still an embryonic industry. One way to grow these unconventional, but nevertheless nutritious materials is to use partly processed petroleum oils. Following work pioneered in France by A. Champagnat and colleagues at the Société Française des Pètroles BP, a number of large oil companies have invested in projects to develop animal feedstuffs from petroleum.

In 1959 the Lavera team commenced a study of the growth of yeasts on pure alkane hydrocarbons. The success of this approach soon led to the replacement of the pure chemicals with 'heavy gas oils', a petroleum distillate which contains about 10 per cent alkane hydrocarbons. The petroleum material is mixed with water containing mineral salts necessary for yeast growth, and air is blown into it. This keeps the immiscible liquids from separating into two layers, and also provides oxygen which the yeast needs to grow when its 'food' is in the form of organic chemicals which do not contain oxygen.

The yeast is fastidious in its 'eating' habits, with the result that the uneaten part of the gas oil can be removed after the process is over. As the yeast consumes a commercially useless constituent, the residue is more valuable when it comes out than when it goes in.

BP has a plant at Lavera capable of producing about 30,000 t/a of its protein feedstuff, Toprina, by this method. It has also built a plant in Italy, in conjunction with an Italian company, with a capacity of 100,000 t/a. This plant uses pure straight-chain alkanes as feedstock. A blow for BP was the decision of the Italian government in spring 1976 to withdraw permission for the company to sell its product for use in Italy. The decision was primarily political; according to the *ad hoc* working group on single-cell protein of the United Nations Protein-Calorie Advisory Group, there were no scientific grounds for the Italian action. Because the Italians remained obdurate, BP decided in April 1978 to close the plant. The Italians' action has probably also discouraged new investment in single-cell protein manufacturing plant. Consequently, it is unlikely that earlier estimates of a world capacity of 1·5 Mt/a by 1980 will be realized.

Natural gas (methane) is another substrate that has been used, in this case for the growth of high-protein bacteria. Shell is currently developing a process which uses methane. ICI, however, found it easier to develop a process in which methane, which is only slightly soluble in water, was first converted to methanol. A smaller-scale process has been developed for utilizing carbohydrates to grow fungi by Ranks Hovis McDougall. This process can be used to make protein from soya-bean residues, that is, the carbohydrate material left after the natural soya-bean protein has been removed and spun into protein fibre materials, such as Courtauld's Kesp, which is already used as a meat substitute in Britain. Apart from the Ranks Hovis McDougall process, the microbial protein products are intended as animal feedstuffs which could replace some of the large quantity of fishmeal currently imported to the U.K. for animal feeding. A number of other companies also have interests in this field. For example, Tate & Lyle is developing processes which will use vegetable or fruit wastes and 'village technology' (cheap plastic fermenters) to produce animal feedstuffs, for indigenous use in developing countries.

One of the major problems in developed countries is not the shortage of foodstuffs, but excess consumption. When people are too addicted to food for their own good health, chemistry can come to their aid by producing tasty, but non-nutritive foodstuffs.

One of the advantages of the cow over the human being is that, with the aid of its stomach bacteria, it can digest cellulose, the structural carbohydrate of plants. For humans there is no goodness in cellulose and, in recent years, chemically modified celluloses have been used in some slimming meals. In some cases, not only is part of the content of the meal harmlessly indigestible, it also swells in the stomach through absorption of liquids, thus giving the eater a feeling of being full. Unfortunately, recent research in psychology suggests that obese people may not respond to a full stomach by stopping eating.

One of the most insidious of foods is common sugar. Some nutritionists claim that it not only causes obesity, but also coronary thrombosis and rotten teeth. However much of this is true – and some of it certainly is – there is a lot of money to be made from non-nutritious sugar substitutes. But it is important that the cure not be worse than the disease. The earliest sugar substitute was saccharine, which was discovered in 1879. About 550 times as sweet as sugar, saccharine has the disadvantage of a bitter aftertaste. This disadvantage does not apply to another group of compounds, the cyclamates. Following nine years of testing in feeding trials, cyclamates were introduced as artificial sweeteners in the early 1950s. They spread rapidly to such a wide variety of foods that it became apparent that some people might consume more cyclamates than the amount which had been assumed as maximum dosage in the feeding trials. Following a number of new trials, and chemical evidence that under certain conditions the cyclamate molecule can rearrange to produce a molecule which has been linked with the induction of cancers, a wave of public feeling broke out against cyclamates in the late 1960s. In 1969–70 they were banned in the U.S. and Britain and food processors in many parts of the world reverted to sugar and saccharine as sweetening agents.

There is no conclusive evidence of the harmfulness of cyclamates, and most experts now believe that the banning of these substances was hasty and ill advised, although, after lengthy consideration, in May 1976, the U.S. Food and Drug Administration refused to lift the ban. Since the last century, when standards of food preparation and preservation were appalling, tests for new products have become more and more stringent, and it is possible that legislators nowadays err too much on the side of safety. However, in some cases, further research has shown that the approach which adopts a very wide margin of safety can be a wise one.

In the late 1940s a substance called agene ($NCl_3$) was used as a flour-maturing agent. In 1947 it was found that dogs fed large quantities of agene-treated flour developed 'running fits'. Although the doses were high, the evidence was sufficient to get agene banned from flour processing. It was later shown that agene reacts with the amino acid, methionine, to form the antimetabolite, methionine sulphoximine, which has serious neurological effects on a number of animal species.

A more recent example of a food additive, the use of which has had to be restricted, is monosodium glutamate. This virtually tasteless amino acid derivative enhances the flavour of foods to which it is added, and was widely used in baby foods. However, in late 1969 and early 1970, U.S. and U.K. food manufacturers placed a voluntary ban on its usage because of its possible action in causing brain damage. This was a wise move for, late in 1971, it was confirmed that large doses could indeed cause damage to a particular part of the brains of young rats and mice.

It is not only man-made chemicals that cause trouble in food. When something natural goes awry, the chemist may be brought in to do some detective work. Perhaps the most famous case in recent years occurred in England in 1960, when 100,000 young turkeys suddenly died from a mysterious ailment which was christened 'turkey x disease'. Eventually the cause was found to be the ground-nut meal on which the turkeys had been fed, which was contaminated with a microorganism called Aspergillus flavus. The active ingredient of the contamination was found to be the chemical, aflatoxin; as little as two hundred-thousandths of a gramme of this substance will kill a day-old duckling.

While it is necessary to keep a watchful eye on the purity of our food, a more important problem is water supply. Water is necessary not only for animals and humans to drink, but also for all plant life. Industrialized societies use water in ever increasing quantities, yet the supplies of potable (i.e., drinkable) water are not endless. Physical science must play an expanding part in obtaining water for mankind, a task that has been left to the hydrological engineer in the past.

There is no shortage of water on the Earth; most of the planet's surface is covered with it. But much of this is either salt or brackish, not suitable for human consumption or even for irrigation, as plants will not tolerate soils with a high salt content. In the hotter countries of the world, the search for new ways to obtain water has been going on for years, and a number of methods are now being tried out.

The two simplest processes are distillation, in which pure water is boiled off as steam from seawater or other impure water, and freezing. The ice formed in the latter process is purer than the water from which it freezes and can of course, be remelted after removal from solution. Both of the processes require large amounts of energy and complex engineering plant. Recently, simpler methods which depend on the widespread physical phenomenon of osmosis have been examined.

Many thin films of complex chemical substances are semipermeable, that is, only certain substances can pass through them. The thin lining between an eggshell and the white of the egg is a semipermeable membrane. If such a membrane is used to separate, say, a strong sugar solution from pure water, the pure water will pass through the membrane to make the sugar solution more dilute, for water molecules, being much smaller than sugar molecules, pass through the membrane with greater ease.

To prevent this from happening, it is necessary to apply pressure to the membrane. Consequently, the osmotic effect can be said to exert pressure, and is often called osmotic pressure. If a higher pressure is applied against the osmotic pressure, reverse osmosis takes place. Water from solution passes through a membrane into pure water on the other side.

Since the late 1950s, work has been going on in laboratories throughout the world to study reverse osmosis and, using films of cellulose acetate as membrane, it is now possible to construct reverse osmosis plants which will purify up to 50,000 gallons of water a day. A great advantage of this method over distillation is that it is also economical to build small plants, producing only say 1,000 gallons a day.

The mechanism of reverse osmosis is complex and is only now beginning to be sufficiently well understood for the design of alternative membrane materials. It seems likely that new membranes will be developed in the near future, although cellulose acetate films, which have been around for some years, are not difficult to make and are sufficiently sturdy for industrial usage. In Appleton, Wisconsin, Consolidated Papers Inc., has been using a portable reverse osmosis plant for several years to clean up its effluent; 90 per cent of the waste water is purified and returned to the plant, leaving behind a concentrated effluent for disposal.

At the moment, reverse osmosis has the edge over electrodialysis, another semipermeable membrane process in which sea water is passed into a semipermeable cell. On either side of the semipermeable membrane is an electrode, one positive and the other negative. When electricity is passed through the apparatus, the salts in the water ionize, and the ions pass through the semipermeable membrane toward the electrodes. As the amount of power required for electrodialysis is proportional to the salinity of the water, the method is only economical where the salt content of the water to be purified is low.

Despite advances in more sophisticated techniques, the large majority of plants currently being built are distillation plants. Partly, this is because the areas of the world that need and can afford desalination plants are often oil-producing countries, where energy is still cheaply available. However, the chemical industry still has a role to play. One problem of distillation plants is the deposition of dissolved salts when the water evaporates. There is a growing market for chemicals to add to the intake water in distillation plants to prevent such deposition, which not only lowers the efficiency of the plant, but shortens its working life.

## Chemistry cures

Having fed, clothed, and sheltered himself, all man needs to enjoy himself is good health. Because of the way we choose to live, we do not often think about our health when it is good, only when it is not. The chemist has played a large part in restoring the sick to health. It would be foolish to suggest that chemistry can provide cure-alls: for example, the scourge of cancer can only infrequently be alleviated by present-day chemotherapy, although some promising drugs are currently on trial.

The most spectacular advances in the field of health care, so far as the chemist is concerned, have been associated with the discovery and development of antibiotics, substances which attack microorganisms and fight off their invasions of the human body. The first of these, discovered in the late

1920s, was penicillin, which is produced by a mould. Intensive study during World War II led to the large-scale production of penicillin by fermentation, as the industrial culture of microorganisms is called.

The penicillin story did not stop here. It was realized, when different fermentation methods were tried, that penicillin was not a single chemical compound, but a group of related substances. Each of these was found to have the same central core to which different 'tails' are attached. In the mid 1950s, when penicillin-resistant strains of microorganisms were causing widespread infection, particularly in hospitals, Sir Ernst Chain, one of the scientists who had worked on the development of penicillin during the war and received a share in a Nobel prize as a result, advised the British company, Beecham, not to search for new antibiotics, but to find ways of modifying the penicillin molecule.

In 1957 a team of research workers at this company isolated the core of penicillin molecule, 6-aminopenicillanic acid (6-APA); two years later this compound was made synthetically by MIT chemist John C. Sheehan, but the synthetic method did not compete economically with 6-APA obtained by fermentation. By 1960 the Beecham team had not only discovered chemical methods for attaching the tail of their choice to the 6-APA core, but had found a semisynthetic penicillin that could beat the drug-resistant strains of microorganisms which had developed. They were able to produce a range of penicillins, all with slightly different properties, so that penicillin still forms a most successful group of drugs available for antibiotic therapy.

While all this was going on, other scientists had not been standing still. At Rutgers University in 1939, a student of Professor Selman Waksman, René Dubos, now professor of pathology at Rockefeller University, had isolated a powerful antimicrobial material, tyrothricin, from the soil-bacteria Bacillus brevis. Tyrothricin was later shown to be composed of two antibiotics (this word was coined in 1942 by Waksman), gramicidin and tyrocidin.

This discovery and the development of penicillin led Waksman, in collaboration with Merck and Co, to undertake a close examination of 10,000 soil cultures in search of antibiotics. In 1943 he isolated streptomycin, for which he received a Nobel prize in 1952. This substance went into production on a pilot plant scale the following year. It was followed, in the next few years, by the discoveries of chloramphenicol, chlortetracycline (aureomycin), neomycin, terramycin (Pfizer scientists screened 100,000 soil samples to find this), and cephaloridine (one of about 600 modifications of a molecule called cephalosporin C). Streptomycin was among the first drugs to make a dramatic attack on tuberculosis; since its discovery a number of synthetic chemicals, such as izoniazid, p-amino-salicylic acid, and cycloserine, have also been introduced as antitubercular or tuberculostatic agents.

Structurally, antibiotics are an odd assortment of molecules. The penicillin core is synthesized *in vivo* from one molecule each of the amino acids, valine and cysteine; gramicidin and tyrocidin are made from amino acids, all of which have the opposite optical configuration to natural amino acids; the tetracyclines, which include aureo-, neo-, and terramycin, are all so called because their basic core is a system made up of four rings of carbon atoms fused together; cephalosporin C was originally called penicillin N, because its core is similar to that of penicillin and it was initially isolated from a culture containing other penicillins.

The modes of activity of antibiotics are also varied. Penicillin, for example, interferes with the synthesis of bacterial cell walls and as mammals do not have similar structures, it is harmless to most humans. Some of the other antibiotics are less selective in their attack. Actinomycin D, for example, reacts with the double helix of DNA, and therefore is injurious to humans as well as bacteria and not very suitable as a therapeutic agent. Chloramphenicol and streptomycin, on the other hand, act by joining on to bacterial ribosomes and prevent them from functioning properly, yet do not disable mammalian ribosomes.

Antibiotics are not the only therapeutic agents discovered since the war. Many new drugs have been synthesized, often on the basis of structures elucidated for drugs obtained from plant sources and known for centuries in impure form. Current theories of how drugs act are, in general, rather hazy, and drug companies find it better to mimic nature than try to design a drug solely on theoretical grounds. Cases in point are analgesic or pain-killing drugs. The best known of these is morphine, obtained from the opium poppy, the therapeutic qualities of which have been known since ancient Babylonian and Egyptian times.

The constitution of the morphine molecule was first worked out in the 1920s (although the exact stereochemistry was not elucidated until later). Since then chemists have been attempting to synthesize variants which will have the analgesic properties of morphine but none of the addictive properties which make its prolonged use so dangerous. So far this work has not been greatly successful, although it has produced useful drugs. In 1964, for example, the British chemist K. W. Bentley, moving in a direction opposite from previous workers in the field, decided to build a larger molecule than morphine. The result was the oripavine drugs which are many times stronger than morphine. They are too strong for human use, but have been used widely in wild game reserves for sedating large animals;

one-thousandth of a gramme of one of these drugs applied to the end of a dart is enough to quiet a two-ton rhinoceros when the dart pierces its hide. It has been found possible with other morphine-related drugs to separate the painkilling activity from the addictive, but it has not yet been possible to produce such a drug without other undesired side-effects, such as hallucinations. However, recent studies of brain biochemistry offer the possibility of rapid advances in this field. During 1973, a group of scientists working at Johns Hopkins University in Baltimore, under Professor Solomon Snyder, discovered the receptor sites in brains with which these opiates react. It has been discovered subsequently that the opiate receptors are distributed unevenly throughout the brain, and are concentrated in areas associated with strong emotions, such as pleasure and rage.

But why should receptors in an animal brain react so strongly with molecules from plants? The most likely answer seemed to be that the molecules had the same shape – at least in part – as an undiscovered natural brain substance. In 1975 the missing substance was isolated by scientists at Aberdeen University, and called enkephaline. By the end of the year, the Aberdeen group had analysed enkephaline and synthesized it.

Enkephaline has actually been found to occur in two forms, both of which are pentapeptides (that is, molecules composed of five amino acids). The amino acid at one end of the molecule is different in one form of enkephaline than in the other. However, both types of enkephaline show spatial similarities, in parts, to parts of the morphine molecule. So it does seem that the effects of morphine on animals are a biological coincidence. On a more practical level, with the structure of enkephaline to hand, it may be possible to design synthetic opiates based on its structure that are more 'natural' in their activity than the opiates we have extracted previously from natural sources.

Another field of synthetic drug chemistry is the production of antidotes to metal poisoning, which involves compounds known as chelates. These are multifunctional organic molecules which can surround a metal ion, forming several chemical bonds with it through the different functional groups. The behaviour of the metal ion is thereby changed, often in a beneficial way. The compounds used to prevent deposition of unwanted substances in desalination plants (see p. 184) are of this type, but their medical use is perhaps more significant. For example, an overdose of some metal may result in the ions of the metal being loosely bound to molecular structures in various organs. Administration of a chelate which can form stronger attachments to the metal will remove it from these organs and help it to pass from the body.

The first synthetic chelate to be used medically was ethylenediamine tetra-acetic acid, a hexafunctional molecule synthesized in 1935 by the German chemist F. Munz. In 1951 this compound, which can wrap around metal atoms to form octahedral complexes, saved the life of a child suffering from lead poisoning. Since then many chelating structures have been designed for alleviation of poisoning by different metals. In particular, chelating agents are useful in removing the ions of various radioactive metals from the human system.

Because of their toxicity, metal compounds have played only a small role in chemotherapy. However, recent work by Professor Barnett Rosenberg of Michigan State University has shown that some metal compounds may be effective in treating cancers. Tests with a platinum-containing drug which he discovered have shown that it is effective against a wide variety of tumours in experimental animals, and clinical trials have shown that it may also cure some human cancers. Rosenberg discovered this drug, the chemical name of which is cis-dichlorodiammineplatinum, by accident. In 1968 he was studying the effects of electrical fields on the bacterium, Escherichia coli. The result of his experiment was that the bacterial cells, instead of reproducing by cell division, just grew bigger, and eventually died. This was not, however, a result of the electrical field; the effect was traced to tiny quantities of platinum compounds which had formed as a result of reactions between the platinum electrodes he was using and the medium in which the bacteria were suspended. Because, basically, a cancer is a group of cells which keeps dividing in defiance of the body's regulatory mechanisms, Rosenberg realized that these platinum compounds might have anti-tumour powers. Although the clinical trials are not complete, it now seems that Rosenberg's accidentally discovered dichloro-compound may be more effective in the fight against cancer than any of the other 300,000 compounds screened for activity during the past fifteen years by the U.S. National Cancer Institute.

## Back to nature

As more becomes known about bodily processes, the accent in the pharmaceutical industry moves from synthetic chemicals to more natural materials, and from the cure of illness to its prevention. The latter is especially true in the case of metabolic diseases caused by genetic defects. Phenylketonuria is a genetic disease in which the ability to use the amino acid phenylalanine properly is absent. Instead, phenylalanine is converted to substances which can cause irreversible brain damage in children. Development of a chemical which reacts with these substances to produce a distinctive colour change has led to a method of spot checking all babies at birth. The poison, if present, appears in the urine, and application to a wet nappy of a small stick or strip of paper impregnated with the appropriate chemical immediately checks whether or not a baby is phenylketonuronic. Those that are can be fed on special diets low in phenylalanine, thus preventing brain damage and enabling them to live healthy, normal lives.

Chemical study of the steroid hormones which control the sexual cycle led to the contraceptive pill – introduced in 1960 – which replaces earlier techniques of contraception with a method which is certain as it uses the body's own mechanism. There are possibilities of side-effects, however, for although the body is not being fed 'unnatural' materials, it is receiving enormous doses of material usually present only in small quantities, and this may have undesirable long-term repercussions.

The latest in the list of natural materials to be taken up by the pharmaceutical industry is the prostaglandin family. Although prostaglandin activity was detected more than thirty years ago, it was not until 1957 that Dr Sune Bergström of the Karolinska Institute in Sweden, using chromatography, spectroscopy, and finance from the Upjohn Company of Kalamazoo, Michigan, isolated pure crystalline prostaglandins. It took another five years to elucidate the structures of the two 20-carbon molecules he had purified. A wide variety of prostaglandins, all with the same basic chemical structure, have now been identified, and the total syntheses of representatives of this group of compounds have been achieved by Professor E. J. Corey and his collaborators in the chemistry department at Harvard University.

All animal cells appear to have the capacity to manufacture prostaglandins, which seem to be involved in the regulation of many bodily functions. This led to hopes that they might be useful in a range of treatments; however, the trend now is toward the manufacture of prostaglandin analogues. For example, prostaglandins have been used extensively to stimulate labour or produce abortion in pregnant women, but a more useful drug appears to be a modified prostaglandin which cannot be broken down by the enzyme system which normally destroys prostaglandins in the body and which prevents the natural compounds from being given orally. Another prostaglandin analogue is used extensively in veterinary practice to synchronize oestrus in cattle which are to be artifically inseminated.

A number of other possibilities are being investigated. For example, some prostaglandins are potent inhibitors of gastric secretion in rats and dogs, but are not active in man either when taken orally or given by intravenous infusion. However, one of the synthetic analogues is stable in the stomach and might be useful in reducing the high acid output of the stomach in cases of peptic ulcer. Another field in which prostaglandin therapy might be effective is the treatment of asthma; some prostaglandins are ten times more effective as bronchial dilators as the commonly used anti-asthma drug isoprenaline. However, there are some side effects which need to be looked at carefully, because it is becoming increasingly appreciated that new drugs must be tested stringently before they can be made generally available.

No better evidence of this can be provided than the thalidomide tragedy. First made in 1953, this synthetic sleeping pill was marketed in Germany in 1957, and later in the U.K. and several other countries but not in the U.S.

Because it provided gentle sleep with no apparent side-effects, thalidomide became widely prescribed for pregnant women. It was not until 1961 that some doctors began to realize that there was a suspicious coincidence of use of this drug and the appearance of hitherto rare deformities in newborn babies, characterized by lack of arms, so that the hands emerged almost directly from the shoulder blades. Thalidomide was hastily withdrawn from the market as the degree of coincidence increased and it became clear that, in some way, the 'harmless' sleeping pill had been responsible for the birth of thousands of deformed babies.

It is to be hoped that such an incident will not recur, although the forward movement of technology leaves us constantly exposed to new risks. Some of these, and the ways in which they can be handled, form the substance of much of Chapter 7.

## The chemistry of art

A few pages back the statement was made that all man needs to enjoy himself are adequate food, clothing, shelter and health. This is probably too much of a generalization: without art and culture it is unlikely that modern man would enjoy himself, however great his health and material wealth.

It has been fashionable in recent years to argue that society is divided into 'two cultures', with scientists on one side of the fence and everybody else on the other, both snarling at each other. Like many facile arguments, this contains some truth, but as the fruits (not all of them palatable) of science spread into every corner of society, there is an awakening of interest in what scientists are doing. This is particularly true in certain areas previously considered the exclusive province of the non-scientist, such as art and archaeology. The British Museum research laboratory celebrated its fiftieth anniversary at the end of 1972, and many museums and art galleries now have laboratories attached. The scientist can help the museum curator and archaeologist as both a detective and a conservationist.

The most famous chemical breakthrough in archaeology is, without doubt, radiocarbon dating, developed in the late 1940s by American chemist Willard Libby. With the increasing knowledge of nuclear transformations and the processes by which one element is converted into another, it became apparent in the late 1930s that nitrogen atoms in the upper atmosphere, exposed to radiation from outer space, were probably transformed into a radioactive isotope of carbon, $^{14}C$.

As carbon is continually recycled in nature, being taken into living organisms through the photosynthetic process and eventually being returned to the atmosphere, Libby proposed that all living organisms would contain a certain proportion of $^{14}C$ in their make-up. When an organism dies, fresh carbon is not taken in. The carbon locked up in the dead molecules gradually loses its complement of $^{14}C$ through radioactive disintegration. Because this disintegration occurs at a fixed rate, it should be possible to tell the age of an object by determining the ratio of radioactive to ordinary carbon in its make-up.

Libby spent several years at the University of Chicago testing his theory. The first step, to see whether carbon compounds freshly released from living organisms did contain $^{14}C$, was shown to be correct through the detection of radiation in methane gas generated from sewage in Baltimore.

By developing a subtle technique which discounted radiation from external sources, Libby was able to measure the small amounts of radiation from $^{14}C$ in organic materials. Through Professor John Wilson of the Oriental Institute at the University of Chicago, Libby and his colleagues obtained samples of organic material of known age from early Egyptian pyramids. Using this, they were able to show that it was possible to date objects by measuring their radiocarbon contents; for this achievement Libby was awarded the 1960 Nobel prize in chemistry.

There are drawbacks to radiocarbon dating. It is based on certain assumptions which may be incorrect about the constancy of upper atmosphere bombardment by radiation. By checking radiocarbon methods against other dating methods, it has now been shown that, as one goes back in time, an increasingly large deviation occurs between dates obtained from radiocarbon studies and those from other methods. A chart of the deviations has now been plotted that goes back 7,500 years. Despite this, the method is still of value. As the mixing of carbon is a world-wide process, two objects of the same age from different parts of the world always give the same radio date. Consequently, an object of known date from one part of the world can be used to date objects of the same age from other parts of the world, even if no independent dating method for these other objects exists. At the moment, radiocarbon dating is causing a gigantic upheaval among scholars of early European civilizations by indicating that these developed several hundred years earlier than was previously believed.

In addition to dating methods based on other radioactive transformations, different methods have been developed in recent years. The constant background radiation which reaches the earth causes damage to molecules. In an object such as an ancient pot, left lying around for thousands of years, this damage, although invisible, is substantial. It is invisible because it usually involves disruption of single atoms, causing minute changes in the material. When such materials are heated above the temperature at which they become red hot, this disruption shows up as luminosity, an effect known as thermoluminescence. The older an object, the more thermoluminescent it is. Although thermoluminescent dates are not precise, the technique can be valuable in detecting fakes. The error – which can be about 10 per cent – makes the method inapplicable to classical archaeological dating, but such an error is insignificant when it comes to determining if an object was made two or 2,000 years ago. Antique objects are rare and command a high price among collectors. This fact has not escaped the notice of various entrepreneurs who have been helping to supply the demand by making 'antiques'; thermoluminescent dating is an excellent way of checking whether the goods offered by these traders are the real thing.

Neutron activation analysis is another versatile technique in the field of historico-scientific discovery. If a material is exposed to neutrons, the nuclei of the elements present become unstable and emit characteristic gamma radiation. This can be used to identify the material's composition. Changes in pigment manufacturing techniques during the past four centuries are clearly shown up by this technique, thus enabling paintings of unknown or dubious provenance to be dated.

These methods can be used to authenticate materials as well as to identify forgeries. The same is true of X-ray fluorescence analysis, in which samples are bombarded with X-rays, and analysed by reference to secondary X-rays emitted subsequently. This technique brought to light an interesting case of non-forgery of some English coinage from the reign of Henry I (1087–1135). It was rumoured at the time that the silver coinage was being debased with tin by the moneyers. X-ray fluorescence has shown this rumour to have been baseless. Unfortunately, the technique came several hundred years too late to vindicate the moneyers who were all assembled at Winchester by the credulous Henry and castrated.

These are only a few examples of the detective work that chemists and physicists are called upon to perform in checking up on the world's art treasures. By combining chemical and physical techniques developed in recent years it is possible to check the authenticity of many different sorts of objects. For example, X-ray diffraction studies of some samples of 'Roman glass' have shown that they have been artificially aged with hydrofluoric acid – a chemical not common in ancient Rome.

On the other hand, the Romans did, until recently, know a few things that we did not. In the British Museum is a Roman glass goblet, the Lycurgus cup, which has a rather striking property. When seen by reflected light it is a sludgy green; when seen by light transmitted through it, it is wine red. Using a minute sample of this cup (seven-thousandths of a gramme), two British scientists found that it had a composition typical of Roman glass, except for the addition of minute percentages of gold and silver. Using this composition of raw materials and trying different methods of melting the mixture to form glass, R. H. Brill, in the 1960s, succeeded in recovering the lost technique of the Lycurgus cup's manufacturer and made glass with the same dichroic effect.

More often, the museum scientist is called upon not to mimic the products of the ancient world, but to prevent their decay. Wooden objects often keep well in water, but once removed from the water and dried out, they crumble away to dust. One solution to this problem is to impregnate the wood with a material that will help to preserve it. This has been done successfully with polymeric materials, the polyethylene glycol waxes, which are water soluble. They can be used to impregnate the wood as the water is removed from it, in a single step. This method has been utilized to preserve wooden objects believed to be more than 50,000 years old, and to treat a seventeenth-century Swedish warship, the *Wasa*, salvaged from Stockholm harbour.

A modified nylon polymer which forms a matt permeable film when applied to objects in alcoholic solution has been used for consolidating flaking paint on oil paintings from Egyptian tombs, as well as for preserving illuminated manuscripts and textiles. Another technique used to restore damaged objects is consolidative reduction. Where metals have been corroded – for example, to halides or carbonates – it is sometimes possible to restore the object by electrically reducing the corroded material back to its original metallic form.

Another method of preservation which may be useful for massive ancient objects is *in situ* polymerization. In 1970, scientists at the Centre for Nuclear Studies in Grenoble, France, impregnated an ancient Egyptian wooden yoke with a liquid resin, which they then irradiated with radiation from cobalt-60. This polymerized the resin to a solid permeating the wood. (It has been suggested that such a technique could be used commercially for upgrading soft timber into hardwood.) Such *in situ* polymerization might provide an answer to the problems of crumbling stonework on such buildings as the Parthenon in Greece, where the damage of centuries of wear has been greatly accelerated in recent years by the effluent from Athenian factories. An alternative method of preservation in this case, of course, would be to eliminate air pollution. But that is a subject for the next chapter.

# *CHAPTER 7*

# CHEMISTRY AND TOMORROW

The two previous chapters have indicated the ubiquity of modern chemistry, not only in the wide sense that all things at a certain level of understanding are molecular, but in the artificial sense of chemicals having been manufactured and applied in myriad ways. As chemicals are so widespread, it is not surprising that some of them are found in the wrong place at the wrong concentration, thus resulting in pollution.

Since the early 1960s, there has been a rapid increase in public awareness of the extent to which man's power over nature can backfire and wreak havoc on his planet. The pollution of air, earth and water are matters of wide public concern, as can be seen from the number of newspaper and magazine articles and hours of TV and radio time given over to their discussion. This examination will be restricted to forms of pollution which are clearly a part of chemistry.

## Pesticides

About one-third of the cereal crops grown throughout the world each year are never harvested by man: disease, weeds and insect pests destroy them. This wastage probably amounts to more than 500 million tonnes of food each year. In some spheres of agriculture the picture is even gloomier. It has been calculated that as much sugar cane and sugar beet are lost as are harvested.

This process has been going on for a long time, and even early civilizations applied their ingenuity to overcoming the wastage. During the early years of this century, the problem of controlling insect pests was passed to the chemist. It was known by then that certain plant extracts were lethal to insects. Nicotine, from tobacco, for example, has been used as a plant spray for many years, and the flowers of pyrethrum and roots of the derris plant (ground down to 'derris dust') are further examples of early, natural chemicals used in pest control. However, it seemed at the time that a simple synthetic chemical, active against a wide range of pests, would be desirable.

Paul Muller, a Swiss chemist working for the firm of Geigy, began to search for such a chemical in the mid 1930s. The search ended in September 1939 with a substance called dichlorodiphenyltrichloroethane, which, for simplicity's sake, was abbreviated to DDT. By the middle of World War II, DDT – which had been discovered in the 1870s but not put to any use – was being used by the armed forces and was also on sale to the public.

During the winter of 1943, in Allied-occupied Naples, DDT showed its power when spraying of humans with a solution of it halted an epidemic of the killer-disease, typhus. DDT has since been widely used throughout the world in the eradication of a variety of pests; it has been estimated that nearly three million tonnes of the substance have been loosed on the world in the past few decades.

This is an enormous quantity of totally artificial chemical to release into the environment, and it has recently been noted that in some ways the use of DDT has been indiscriminate. It would be wrong to say that DDT has caused more harm than good. DDT has done and, under controlled conditions, still can do an immense amount of good, but it is possible that more thought would have allowed its deleterious side effects to be seen earlier.

Professor Barry Commoner describes a personal experience from World War II, when he was a U.S. Navy project officer working on the development of DDT spraying by aircraft, 'which proved to be of great importance in the Pacific island battles by protecting the first wave of attackers from serious insect-borne diseases'. A request came to the Navy from a rocket experimental station on an island off New Jersey for help in destroying a plague of flies which was interfering with the experimental work. 'We sprayed the island and, inevitably, some of the surrounding waters with DDT. Within a few hours the flies were dead, and the rocketeers went about their work with renewed vigour. But a week later they were on the telephone again. A mysterious epidemic had littered the beach with tones of decaying fish – which had attracted vast swarms of flies from the mainland. This is how we learned that DDT kills fish.'[1]

Shortly after the development of DDT, a new insecticidal compound, BHC (benzene hexachloride), was discovered almost simultaneously in England and France. In the late 1940s chlordane, aldrin and dieldrin were discovered by Julius Hyman and his colleagues, working first at the Velsicol Corporation and later in Hyman's own company. Together with heptachlor and endosulphan, these five are the best known of the organochlorine insecticides, so-called because they are all organic molecules (and

1. Barry Commoner, *Science and Survival,* Ballantine Books, New York, 1970, pp. 25–26.

contain at least one modified benzene ring), and all contain several atoms of chlorine.

It is reasonable to treat them as a group, for they tend to be long-lived substances which accumulate in living tissues. For example, the fish kill described by Commoner was trivial compared with the accident in the river Rhine in June 1969. Two barrels of endosulphan, accidentally dumped in the river in Germany, are estimated to have killed forty million fish. In addition, the Dutch, who take drinking water from the Rhine, were forced to fall back on emergency water supplies while the 'mysterious epidemic' – the cause was not discovered for several days – was rampant.

Birds have been the worst sufferers from organochlorine insecticide poisoning. DDT acts on the liver, probably interfering with body levels of the hormones concerned with reproduction. This shows up in the production of eggs with exceptionally thin shells which break before the chicks are ready to hatch.

It is not possible to say what the long-term effect on man may be. In 1969 the U.S. Food and Drug Administration condemned 700,000 coho salmon from Lake Michigan because of their DDT content. Since 1972, there has been a total ban on the use of DDT in the United States. However, Nobel prizewinner Norman Borlaug has pointed out that it would be criminal to ban its use everywhere. In the underdeveloped countries of the world, cheap effective insecticides are still an urgent necessity. There is equally no excuse for the advanced countries to continue indiscriminate use of organochlorines when there are alternatives, albeit more costly ones.

Following work on organophosphorus nerve gases (see Chapter 4), a different class of insecticides, the organophosphorus compounds, was developed. These have the advantage of not being very stable chemically. They remain in soil, for example, only a few months before being degraded to harmless products, rather than the years taken by some of the organochlorines. The first world-wide success of the organophosphorus compounds was parathion, a wide spectrum insecticide. Perhaps more important from the methodological viewpoint was schradan, which proved to be the first 'systemic' insecticide. It is readily taken up by plants and distributed throughout their vascular system, making the whole of the plant resistant to insect attack. When applied directly to insects, systemic insecticides are not usually effective, but once they have been absorbed by the plant, and the feeding insect swallows a mouthful of insecticide together with some plant sap, the lethal properties become apparent.

In addition, chemists have synthesized a number of other types of compounds – chemosterilants, which make the insects lay infertile eggs;

antifeeding compounds, some of which are organometallic rather than straightforward organic compounds, and which work by making the insect give up eating and starve itself to death; and the sex attractants (pheromones), whose use in pest control has already been mentioned.

Early in 1975, an American company, Zoecon Corporation, which has been working for several years on the development of biologically-oriented pesticides, received permission from the United States' Environmental Protection Agency to market the first juvenile hormone analogue pesticide. Altosid SR-10, which was approved for the control of mosquitoes by prevention of larval metamorphosis, could be the first of a series of synthetic growth regulators. However, in August 1975, scientific evidence appeared to show that there may be problems associated with the use of some bioanalogues. Jeremy McNeil of Laval University in Quebec showed that when a particular type of potato aphid was treated with juvenile hormone analogues, the analogues had a more harmful effect on one of the aphid's natural parasites than they did on the aphid itself. Effects such as this, if widespread, could severely limit the utility of the bioanalogue approach to pest control.

Renewed interest has recently been shown in some of the more complex natural insecticides from plants, the structures and laboratory syntheses of which have been achieved in recent years. Pyrethrin, still obtained from plants related to the chrysanthemum and grown mainly in Kenya, is widely used in aerosol cans of flykiller, as well as larger-scale applications because it has a potent, but short-lived effect. Work at Rothamsted Research Station in England has led to the development of synthetic analogues of pyrethrin. Resmethrin, discovered in 1967, is being produced in the United States and Japan for use on indoor insect pests and house plants. Another analogue – more potent because it is less resistant to degradation by sunlight – is NRDC-143, discovered in 1973. This appears to be up to 100 times as effective as DDT, and several times as effective as the organic phosphates, and is presently undergoing field trials in Britain and America. If the trials prove satisfactory it has been estimated that up to 20,000 tonnes/annum of this compound could be used in Britain by 1980.

There is no doubt that research on insecticides will continue, much of it motivated by the desire to overcome the defects of the early synthetics. The discovery of DDT in Antarctic ice and other remote regions has brought home the fact that a chemical accident can no longer be confined; it soon reaches everyone's back garden.

## A breath of not-so-fresh air

The Earth is a large system composed of interlocking cycles. As Libby's development of carbon-14 dating showed, there is continual and complete mixing of the planet's air. Similarly, through the water cycle, in which seawater is evaporated by solar heat, the vapour moved as clouds, and then precipitated as rain in other places, the waters of the Earth are mixed – and to the seas and rivers are added substances picked up from soil as rainwater trickles through it.

Since the industrial revolution, man has poured massive quantities of chemicals into the air and water, resulting in measurable compositional changes. The industrial revolution was built on energy supplied by the burning of fossil fuels, first coal, later oil and natural gas. The products of this combustion are mainly carbon dioxide and water. It has been estimated that, since 1850, the concentration of carbon dioxide in the atmosphere has increased from 290 to 330 parts per million. As much as a quarter of this increase may have occurred during the past decade. If present trends continue, the atmospheric concentration could double by 2020. Not all, but a substantial part of this increase can be attributed to the burning of fossil fuels. Whatever the source, the increase has concerned a number of scientists who believe that the effect of this extra carbon dioxide could be to change the Earth's climate.

Much of the energy which the Earth receives from the sun is taken up by soil, then radiated back into space as energy of a different wavelength. Carbon dioxide can absorb this terrestrial radiation, cutting down the amount returned to space. If the carbon dioxide content increases, so may the mean temperature of the Earth, and a rise of only a few degrees in world temperatures would melt sufficient of the polar ice caps to submerge many of the world's cities.

Such an effect might be countered by the emissions of grit and dust which accompany the burning of fossil fuels (but which are gradually being legislated to lower and lower emission levels). These may be increasing the turbidity of the atmosphere and cutting down the amount of radiation which reaches the Earth – hence having a cooling effect, in opposition to carbon dioxide's heating effect. It is important to bear in mind that both these processes may be of only secondary importance when compared to continuing natural climatic changes. The eruption of the volcanic island Krakatoa in 1883 spewed more particulate matter into the atmosphere than man has done throughout his entire history. The Earth has passed through a series of ice ages, alternating with warm spells, and the forces which dictate these long-term fluctuations are probably more powerful than any of the man-made contributions. However, the problem of gaseous emissions as a result of human activity should not be ignored.

Many fossil fuels are contaminated with sulphur-containing compounds, the burning of which produces gaseous sulphur dioxide. This is an unpleasant air pollutant, but can be controlled. In general, sulphur dioxide seems to be a localized irritant, associated with the airspace over heavily industrialized areas. It has even been suggested that it may have a beneficial effect. Heavy agricultural cropping of land, made possible by widespread application of fertilizers, can lead to depletion of minerals in soil. Any sulphur deficit produced in this way may be made up at present by absorption of sulphur dioxide from the air. Cleaning up sulphurous emissions might lead to sulphur starvation in crops. However, as agricultural scientists are now more concerned with designing fertilizers individually for different areas and making up mineral deficiencies on a local basis than with selling straightforward NPK fertilizers, the worry of sulphur starvation should not be used as an excuse for not cleaning up the air.

One factor which could lead to an abrupt decline in sulphur dioxide emissions is an increase in the price of elemental sulphur. All the scientific and technological work needed to remove sulphur dioxide from smokestacks has been accomplished, but the processes, which lead to sulphur recovery, are not economically competitive with sulphur obtained from mines. If they ever become so, the laws of economics, if nothing else, will end this source of pollution.

The major increase in the burning of fossil fuels during the twentieth century is a result of the spread of the motor car. In addition to emitting carbon dioxide and its more poisonous relative, carbon monoxide, cars and lorries also belch out oxides of nitrogen and partly burned hydrocarbons. Under some conditions, notably those in Los Angeles, this mixture, when provided with energy from sunlight, can undergo complex photochemical and free radical reactions to produce compounds irritant to living organisms. As well as peroxyacyl nitrates, acrolein and formaldehyde, all of which cause eye irritation, the reactions involved produce ozone, a molecule composed of three linked atoms of oxygen, unlike the normal diatomic oxygen molecule. Ozone is highly reactive toward substances such as rubber, and is the cause of abnormally high levels of rubber cracking in the Los Angeles area.

To make gasoline burn more efficiently in high-compression engines, a number of chemicals are added to it. The most common additive is the organometallic compound, tetraethyl lead. Car exhausts emit lead and organolead compounds from combustion of fuel containing this additive, and it has recently been suggested that we may be contaminating the environment with lethal quantities of lead because of its use as an additive. This is a controversial subject on which the scientific evidence is still insufficiently clear to say whether or not the lead is causing harm. Because of the possible dangers, a number of countries have introduced legislative programmes to reduce the lead content of petrol. However, as unleaded petrol gives poorer mileage per gallon than leaded petrol, escalation of oil prices and pressure from the manufacturers of lead-based additives have halted or slowed down some of these programmes.

Man's usage of metals is a major cause of concern. Most metals are toxic to living organisms in all but the smallest quantities. Yet industrial activity is leading to the pollution of land and sea not only with lead, but also mercury, cadmium, copper, chromium and zinc. It has been estimated that atmospheric transport is dumping about 200,000 tons of lead into the oceans each year, most of it derived from the tetraethyl lead in gasoline. Many of these metals are concentrated by molluscs, which thus become poisonous to whatever feeds on them.

There is no doubt that mercury in some forms can be lethal. In Japan, between 1953 and 1963, more than 100 cases of serious nervous illness leading to severe debility or death were shown to be directly attributable to consumption of oysters which had picked up methylmercury discharged by a chemical company into Minamata Bay. In late 1971, after a long legal wrangle, the company responsible had to pay substantial compensation to victims of mercury poisoning.

In addition to metals, the most persistent pollutants of the hydrosphere are the organochlorine pesticides and other organochlorine compounds, notably the polychlorinated biphenyls (PCBs) and waste products from the manufacture of polyvinyl chloride. As the sea is, in many respects, our last resource, not only for new supplies of minerals but also for food, pollution by materials known to concentrate in food chains should be minimized. Nevertheless, during July 1971 the tanker *Stella Maris* took 600 tons of chlorinated hydrocarbon waste from a Dutch factory out into the Atlantic with the intention of dumping it. Only an immediate international outcry prevented this.

Most industrial activity is more responsible. In late 1970 and early 1971, Monsanto Chemicals, sole U.K. and U.S. manufacturer of PCBs, restricted sale of these products to carefully defined uses where they could not cause pollution. Subsequent measurements have shown a decline in PCB levels in the environment. It now looks as if PCB manufacture may soon cease entirely, their applications being taken over by specially developed siloxanes (polymers related to the silicones).

## Dirty ol' man river

The problems of marine pollution differ in many respects from those of river pollution, particularly because of the high utilization of river water for drinking. Compounds which will slowly break down to harmless materials in the sea may be dangerous in rivers, because they will be ingested by land animals before they have decomposed. Excessive applications of fertilizer – which may result in the destruction of soil structure – lead to inorganic nitrates running off into rivers. These can be converted microbiologically to nitrites, which are dangerous to human health, notably to babies, because they interfere with the respiratory process. The increase in nitrate levels can also lead to massive build-up of algae, which deoxygenate the rivers and make them uninhabitable for fish. The problem is also aggravated by the phosphates poured into rivers as a result of household detergent usage.

Synthetic detergents are a good example of how modern technology can go wrong. From the 1940s onward they began to replace soap as the major household cleaning agent; the basic ingredients in these detergents were branched-chain modified hydrocarbons derived from petroleum (soap is made from animal or vegetable fats). From the householder's viewpoint the major difference between the two is that detergents work with equal efficiency in hard or soft water, and do not form the insoluble scums characteristic of soap. To industry, until recently, reliance on a petrochemical raw material has been far preferable to reliance on an animal- or plant-derived material subject to agricultural supply and price fluctuations. What was overlooked was that microorganisms, which can metabolize animal and vegetable materials, cannot break down branched-chain hydrocarbons. Rivers began to foam from the waste detergent poured into them. By 1960 alone, nearly two million tons a year of synthetic detergents were being used and thrown away. Tap-water, culled from the rivers and purified by conventional processes which did not remove the detergents, came out of taps in some areas with a 'head' on it like a glass of beer.

In 1965 detergent manufacturers voluntarily gave up the use of branched-chain hydrocarbons in their products and replaced them with straight-chain hydrocarbons, which microorganisms can break down. Since that time the spotlight has turned on the phosphates which also go into detergents (as emulsifiers and water-softening agents) because of the part they play in altering the ecology of rivers. Already, phosphate detergents are banned in Indiana, Minnesota and New York states.

Detergents have also been the *bête noire* of another world-famous incident. Part of the growing dependence of man on fossil fuels has led to a large increase in the size of oil tankers. In March 1967 the tanker *Torrey Canyon* crashed on the Seven Stones reef off the south-west coast of England. One hundred thousand tons of oil was set adrift on the sea, headed toward the beaches of England and France. The English decided to disperse the oil by massive spraying of detergents. When the exercise was studied in detail later, it was found that while the oil had been responsible for a substantial number of bird deaths, far greater havoc had been wreaked on marine life by the toxic effects of the detergents.

Since the *Torrey Canyon* disaster, a number of chemical companies have developed detergents much less toxic to marine life, and have also pioneered other methods of clearing up oil spills–for example, by soaking up the oil with a carpet of polyurethane beads or flakes. But it is small credit to cure problems when prevention would have been possible with a little foresight. The shortcomings of modern technology were highlighted again in March 1978, when the tanker *Amoco Cadiz* was wrecked near the French coast, causing serious pollution of neighbouring beaches.

### Too energetic by half

The pollution caused by both oil spills and the burning of fossil fuels raises an important question: Should we continue relying on these sources for our energy supply? Estimates show that at the present rate of usage, the world's known reserves of oil and natural gas will last for only about forty years. If we assume a 5 per cent annual growth rate in consumption, the lifetime of these reserves is roughly halved. Coal reserves are more substantial. At the present rate of use, known reserves would last for more than a thousand years. However, even as low an annual growth rate in consumption as 2 per cent reduces the lifetime of these reserves to less than 200 years. And, if oil and gas run out – or even become much more expensive – the rate of coal consumption will probably increase far faster than 2 per cent per annum.

There are probably undiscovered reserves of fossil fuels, but the rate at which new discoveries are made is decreasing markedly. There are also relatively untapped reserves, such as oil shales, which could increase in importance. However, in all cases, supplies are limited and, what is often overlooked, these materials can provide valuable raw materials for the manufacture of chemical products, such as polymers. To burn them is, largely, to waste them.

If we stop burning the fossil fuels, where will the world's energy come from? Overlooking the fact that, despite government-sponsored and well-publicized conservation campaigns, the developed countries waste enormous amounts of energy (in the U.S., per capita energy consumption is 100 times that needed to sustain life), what are the alternatives?

In parts of the world where the geography is favourable, water power has been harnessed to generate hydroelectricity. It has been suggested that the power of sea tides could be used to generate electricity in a similar way in other parts of the globe. The British government announced, in May 1976, a two-year $5m research programme on wave power. Because of its extensive coastline, Britain is one of the best-situated countries in the world for extracting energy from waves. And, in the United States, the prospect of commercially harnessing geothermal energy is under careful scrutiny. But these methods suffer severe geographical limitations. Ideally, since there is so much of it, we could try to harness solar energy or – rather than harness it – store it and concentrate it, so that it will be there when we want it. To do this, however, a number of major technical problems must be overcome.

In the meantime, there is nuclear energy. According to Hans Bethe, an American Nobel prizewinner, nuclear fission is the only major non-fossil power source the U.S. can rely on for the rest of this century and probably for some time afterward.[2] An outline of the civil uses of this energy from decomposing atomic nuclei appeared in Chapter 6 where it was mentioned that concern about the environment could slow down, or even stop, the growth of programmes in this sphere. Why should this be so? Unlike fossil fuels, fissionable nuclei decompose at a fixed rate regardless of whether or not man makes use of them. What can be harmful about harnessing the energy of this decomposition process?

The fault in this argument is that, under ordinary circumstances, the radiation given off by radioactive elements as they occur in rocks is dilute; when a nuclear fuel is made, radioactivity is concentrated. It is also probable that this concentrated radioactivity will be transported to a site near large-scale human habitation, as it is economically advantageous to have power-generation facilities as close to the power user as possible. And radioactivity is extremely harmful to living matter. In addition, the new programmes to develop fast breeder reactors rely on decomposition of radioactive elements into new radioactive materials that would not have been produced under 'natural' conditions; man is not just using radioactivity in rocks, he is generating additional radioactivity.

The nuclear lobby argues that nuclear power stations are safe, with an

2. H. A. Bethe, 'The necessity of fission power', *Scientific American*, ccxxxiv, 1976, p.21.

extremely low likelihood of accident. When the type of accident involved may mean a nuclear explosion near a large city (which could happen if a fast breeder reactor went disastrously awry), the risks, some people argue, are not worth taking. One of the earliest nuclear power plants was built at Windscale in the north of England. When an accident occurred at this plant in 1957, a substantial amount of radiation was released over the neighbouring countryside. There have been other accidents, varying in degree and kind. Although it did not cause any release of radioactive material, the fire at Browns Ferry nuclear plant, Athens, Alabama, shows how sensitive such plants can be to 'human error'. Just after noon on 22nd March 1975, two electricians in the cable spreading room beneath the control room at Browns Ferry were using a candle to check for air leaks into the secondary containment area. The naked flame ignited the polymeric sealant in which the cables were set, the cables then caught fire, and flames were sucked into the reactor building. The whole plant, generating 2,200 MWe, had to be closed down – which was done rapidly, although not straightforwardly, for the fire had destroyed some of the cables needed for the automatic shutdown procedure.

There are other worries. Even if nuclear power plants are built on uninhabited land, away from major centres of civilization (which means increased costs for transporting the energy generated), they will eventually wear out. Once a nuclear power station is obsolete, the land cannot be used for anything else because of the high radiation levels. As plants get bigger, the problem of disposing of the radioactive waste increases. In the United States alone there are at least 75 million gallons of highly radioactive waste stored in tanks, because no other satisfactory method has been found of disposing of it. In April 1973, at Hanford Reservation, in southeast Washington State, where 75 per cent of the United States' nuclear wastes are stored, one of the steel-lined tanks sprang a leak. By the time the leak was discovered in early June, 115,000 gallons of toxic waste had escaped.

This was not Hanford's first waste disposal problem. A year earlier the U.S. Congress had had to give the U.S. Atomic Energy Commission nearly $2 million to establish the world's first plutonium mine. Some of the wastes at Hanford had been dumped over a long period into concrete-sided pits with unlined bottoms, from which they trickled down into the earth. The AEC, to its dismay, discovered that the earth acted like a giant chromatographic column, separating the different elements into bands concentrated at different levels. At one level, there was so much plutonium that heavy rainfall could have set a nuclear chain reaction going. (Water acts as a 'moderator' in many nuclear reactors, slowing down fast neutrons so that they are more likely to be absorbed by adjacent fissionable nuclei, which then break apart. It thus helps to sustain the chain reaction.)

The idea of a nuclear reactor establishing itself by accident may seem far-fetched. However, we know it can happen – because it did, nearly 2,000 million years ago. In Gabon, in western Africa, are the fossil remains of a natural fission reactor. French scientists who have examined the site estimate that the reactor, produced by geological anomalies, operated for about 700,000 years, consumed 800 tons of uranium metal, and released enormous quantities of energy.

A process which is effective in some ways in disposing of wastes is to seal up the radioactivity by mixing the active wastes with silicates and fusing them into glass. This localizes the radioactivity, but one still has to put the glass somewhere out of the reach of man and animals. There are already documented cases of people dying because they have found discarded radioactive materials and handled them carelessly, not knowing what they were. Death from radiation sickness is usually extremely unpleasant, and it seems clear that much hard thinking will be needed to discover a satisfactory method of ridding ourselves of nuclear waste.

At the moment everybody's ultimate rubbish bin, the world's oceans, are receiving their share of radioactive waste, but this may stop as concern mounts about the way we treat our massive but finite seas. It has already been shown that it is no good relying on great dilution to render noxious materials harmless, because living processes may reverse this trend. A study on water waste from the Hanford nuclear reactor, which goes into the Columbia river, has shown that the radioactivity can be concentrated one-millionfold by river plankton.

Despite the associated problems, nuclear energy will probably continue to grow in scale. The United States Energy Research and Development Administration, which started work at the beginning of 1975, and is now part of the U.S. Energy Department, spent about half of its first year's $3,700 million budget on nuclear energy development. During the first half of 1977, nuclear power produced 12 per cent of U.S. electricity, thus surpassing hydroelectric power. However, nuclear energy can never solve the energy crisis alone. As Mihajlo Mesarovic and Eduard Pestel showed in their report to the Club of Rome,[3] for all primary energy to be nuclear in 100 years time, we would have to build four reactors per week throughout the whole of the intervening period. And, because reactors wear out, the world would eventually need two new reactors every day, just as replacements.

There are about 250 nuclear reactors now producing power throughout

3. Mihajlo Mesarovic and Eduard Pestal, *Mankind at the Turning Point* (The Second Report to the Club of Rome), Dutton, New York, 1974.

the world, and it has been estimated that there will be more than 400 by 1980. One limiting factor in the growth of nuclear power could be the energy required to construct the reactors. Eventually, of course, one gets far more energy back from a reactor than goes into its construction. Nevertheless, very rapid expansion of nuclear power could create its own energy crisis by draining existing energy supplies that are needed elsewhere.

There is, of course, another form of nuclear energy which could supply man's needs, without creating such insoluble problems – provided that the problem of developing an effective process for harnessing it does not prove insoluble. Nuclear reactors capture the energy released when large, unstable nuclei break apart. The converse of this fission process is fusion: the combination of two small nuclei to form a larger one.

Experiments to harness fusion power have been undertaken, without substantial success, for many years. However, scientists in the field are currently optimistic. The major problem in developing a fusion process has been the containment of a very hot mass of ionized gas, called a plasma, which has to be generated and maintained in order to overcome the energy activation barrier to the fusion of nuclei. A plasma will not only melt any material with which it comes into contact, but will also cool down and cease to provide the energy required to initiate fusion. Consequently, containment (in 'magnetic bottles') is one of the major unsolved problems in the development of fusion processes.

An alternative, currently also being explored, is to initiate fusion in a pellet of material – the most likely fusion fuel is a mixture of deuterium and tritium – with a laser beam. By focussing the beam's energy very precisely, it might be possible to initiate the fusion reaction. Making either type of system work still has to be done, and some scientists, such as Edward Teller, the American who did much of the developmental work on the hydrogen bomb and who has been a protagonist of civil engineering with nuclear explosives, are sceptical about the likelihood of a satisfactory process being developed during this century.

In the meantime, there is an excellent fusion process in operation, of which we might take greater advantage – namely, the Sun. If we could develop new ways of concentrating and storing solar energy, we could probably do without any other energy source in the long-term. In principle, it could be done now. Solar farms, perhaps in desert regions where there will be no competition from agriculture for the use of land, could use the Sun's energy to decompose water into hydrogen and oxygen, either chemically or biochemically. Hydrogen could then become our basic fuel. With more money and effort, it is probable that much better systems

– probably using more complex chemicals, perhaps even generating synthetic petroleum – could be developed. There is still a lot of engineering development to be done but, as Mesarovich and Pestel noted: 'If governments are willing to fund research as generously as they funded nuclear energy, space exploration and supersonic aircraft, operational feasibility and economic reality should be achieved rather quickly.'[4] A Solar Energy Research Institute was established in the U.S. in 1974 and the budget for research in this field is now about $300 million a year, compared to about $100,000 a year between 1950 and 1970. On the domestic front, it has been estimated that solar heating systems are being fitted in U.S. homes at the rate of 80,000 a year.

## If all the world were plastic

As in most social areas, the question of energy supply involves balance. If we want energy, it must come from somewhere. Apart from a few areas of the world that rely on hydroelectricity, the energy we generate pollutes either air or earth. Fossil fuels used for energy not only pollute the atmosphere, they destroy intermediates that could be of use to the petrochemical industry as future raw materials.

Yet if we save the fossil fuels for raw material use they will still, unless we are careful, become pollutants. One of the major groups of end-products from petrochemical operations are plastics, many of which, used in packaging and similar ephemeral purposes, end up as litter. In the advanced countries we not only consume vastly more energy than we need to survive, we also consume more matter than we really use. The result, in both cases, is pollution.

In the United States the average annual per capita production of rubbish is 1,000 lb; in the United Kingdom it is 650 lb. We are producing more rubbish than ever before, and its composition is quite different from that of the rubbish of our grandfathers. An investigation in London showed that, between 1888 and 1967, the percentage of fine dust and cinder in household refuse fell from 82 to 19 per cent. The amount of paper waste increased from 0 to 34 per cent, while the metal and glass content increased from 0·4 per cent and 1·3 per cent respectively, to nearly 11 per cent each.

As time goes by, the composition will change even more. Plastics are replacing paper, metal and glass materials as containers for a variety of goods. This creates difficulties in refuse disposal; incineration of some

4. Ibid, pp. 140–1.

plastics, for example, releases hydrochloric acid gas into the atmosphere. A greater problem is the disposal of such materials outside the refuse-collecting and disposal system. Paper litter left in beauty spots will disintegrate under the effects of climate, but many plastics do not degrade. They remain as unsightly messes, unpleasant to look at and dangerous to some animals, which have been known to die after getting plastic litter caught in their throats.

There has been research in recent years to design plastics which degrade after a few months. The principle is simple: either an additive or integral part of the polymer molecule is made photoreactive, so that the polymer breaks down under the action of light to fine dust and, in the process, to molecular sizes amenable to biodegradation. There are problems. How do you calculate how long a container will be in use before it is thrown away? If consumers who do not consume as fast as the average find their half-full plastic containers of jam, sugar, or corn oil disintegrating in the home, they will be less than satisfied. A study published in 1974 by the Battelle Institute in Frankfurt, for the German government, concluded that degradable plastics had been vastly overrated, and drew attention to the possible dangers of poisonous degradation products being leached into water supplies.

As with the question of energy supply, many problems of pollution in the developed countries could be solved if people consumed less. To bring this about requires a revolution in society rather than in chemistry. However, any such revolution needs to be based on the best knowledge available from all disciplines, both scientific and social. So it is relevant to look briefly at the scientists' role in society, and the ways in which it may be changing.

## The role of the scientist

Analysing the scientist and his behaviour has become a popular activity among the postwar generation of sociologists and psychologists. One result of this activity has been the promulgation of a collection of platitudes about scientists. Possibly the most basic of these is that, of all the scientists who have ever lived, 90 per cent are alive today. Despite the prevalence of this statement, it is never doubted that all those alive, in some essential way, work in exactly the same manner as the 10 per cent who went before them, back to the beginning of recorded time. That is, they all employ 'the scientific method'. Yet, as the Nobel prizewinning biologist Sir Peter Medawar pointed out in a lecture at the University of Pennsylvania in April 1968: 'Most scientists receive no tuition in scientific method, but those who have been instructed perform no better as scientists than those who have not. Of what other branch of learning can it be said that it gives its proficients no advantage; that it need not be taught or, if taught, need not be learned?'[5]

This might be sufficient to make us ask whether the 'scientific method' has any reality other than that of an intellectual construct. Yet the common account of the scientific method has appended to it several corollaries which, until recently, all scientists have believed, or at least paid lip-service to. The most important, perhaps, is that a scientific discovery is neither good nor bad; it is only the use to which it is put that can have either of these qualities. Consequently, utilization of his discoveries is not the legitimate concern of the scientist.

This ethical position was probably adequate for the early scientists, most of them amateurs, the cost of their research supported from their own pockets or from those of philanthropic patrons, and their discoveries only rarely put to use in their own lifetimes. The position of the scientist now is much different. Most scientists are paid by governments or private companies which employ them to make discoveries primarily for material gain, not for self-gratification. If a scientist works for a company whose main object is the manufacture of weapons systems, he cannot say that any discovery he makes is morally neutral. He knows perfectly well that the end to which his discoveries will be put is military and that, if his discoveries are consistently useless in this respect, the company will probably fire him.

At the time when he and his colleagues were spraying DDT during the war, Barry Commoner says they were 'fat, dumb and happy'. He and a growing number of other scientists are no longer prepared to stay that way. It has been suggested that, from their efforts, a new field, 'critical science', is emerging. This is a far-fetched claim; it seems more realistic to say that science is catching up with the world it has created. Science, like all intellectual disciplines, is organic: it grows and develops. To insist that it stick rigidly to the ground rules it obeyed in the seventeenth century is a nonsensical position.

One of the most stimulating theories about the growth and development of science is that put forward by Thomas S. Kuhn of the University of Chicago.[6] Kuhn suggests that sciences develop by undergoing periodic revolutions in which a major element of theory is changed. When an old theory no longer fits all the evidence, scientists struggle along with it,

5. P. B. Medawars, 'Induction and Intuition in Scientific Thought', Memoirs of the American Philosophical Society, vol. 75, 1969.
6. Thomas S. Kuhn, *The Structure of Scientific Revolutions,* 2nd ed., University of Chicago Press, 1970.

propping it up around the corners, until eventually somebody produces a new and better theory.

Usually there is an old guard who bitterly defend their theory against the onslaught of the new theorists. Full acceptance of the new theory often must wait until the old guard die. When a theory is put forward, it opens up new vistas of possible experimental work to verify and expand it – and eventually produce the contrary results that lead to cracks in its façade.

What may now be happening is that science itself – or our idea of it – is undergoing a Kuhnian revolution. Emergence of the youth counterculture which rejects the benefits of 'progress', the growing belief among economists that there is no direct relationship between scientific discovery and greater wealth, and the increasing realization among scientists themselves that they are working in a small, virtually closed system – the Earth – and are themselves a part of their experiments, are all modifying the way in which scientists look at and go about their work. This is not to say that an actual experiment is carried out according to rules of observation different from those used by Joseph Priestley, Sir Humphry Davy, or any of the other great names of early chemistry. The changes are in the planning of which experiments are to be done and the terms in which the results are analysed.

In many cases this means that scientists can no longer live according to the disciplines in which they were trained. The chemist who wishes to design a new material must collaborate with biologists and microbiologists to check its effect on living systems, and with engineers to determine what design features are most needed. The work of meteorologists and hydrologists in plotting currents of air and water is of key importance to the chemist and chemical engineer concerned with the siting of new plants which will produce effluent to be disposed of.

The increased need for interactions among scientists of different disciplines may also reflect an even greater challenge: reintegration of science with the other leading concerns of society. Scientists and technologists must work together with psychologists, sociologists, urban planners, lawyers and many others with whom they may be unfamiliar. The pressing dilemma is no longer how to achieve mastery over nature, but how to use that mastery, or, more important, how to assess whether or not it really is mastery.

At one time, if scientific or technological innovation made it possible to make a new product, or offer a new service at a profit, then it was done. This was called 'progress'. Now we must learn to choose between alternative technologies. This may, for example, mean finding a basis for weighing the advantages to a select group of businessmen of time saved in supersonic air travel against the disadvantages to people on the ground who suffer excess noise nuisance as a result of this form of transportation. Scientists may argue that this is purely a social problem and that, except in his role as private citizen, the scientist has no more right to take a stand on the issue than anyone else. Others, such as Commoner, would argue that he may have no more right to take a stand, but his education and training endow him with a duty to inform his fellow citizens, so that the decisions they make are informed social choices.

Pollution is not really a scientific problem. Many of the substances which pollute our environment could be cleaned up at source if it were economically advantageous. The technical methods for controlling many forms of pollution have already been worked out; to implement them requires changes in legislation which, in the western world, spring ultimately from the people. This can only happen if the people are informed.

Rather than clean up their effluents and accept a cut in profits, some companies will mislead the public with stories of having to close factories and create unemployment if they are no longer permitted to pollute the atmosphere. Yet it is surprising how many companies have managed to survive, once legislation has forced them to act. In fact, according to a study carried out by the Center for Policy Alternatives at the Massachusetts Institute of Technology environmental regulations are often more successful at spurring innovation than direct government efforts to stimulate new product development. Further, although the regulations provide the stimulus for change, the new products that result are often far better than the regulations required.

Barry Commoner has written:

> Scientific method cannot determine whether the proponents of urban superhighways or those who complain about the resultant smog are in the right, or whether the benefits of nuclear tests to the national interest outweigh the hazards of fallout. No scientific principle can tell us how to make the choice, which may sometimes be forced upon us by the insecticide problem, between the shade of the elm tree and the song of the robin.[7]

But as science is forced increasingly into the public arena, partly as a result of rapidly increasing costs of research (which mean that ever larger subventions of public money are needed to keep it going), and partly because of the widespread and rapid effects of scientific innovation, the public has a right to call upon scientists for an explanation of what they are doing

7. Commoner, *Science and Survival*, p. 121

and why they are doing it. As the obligation clearly lies with the scientist, it is up to him to keep on explaining until the public has the information it wants in a form it can grasp. It is unethical of a scientist to rely on technological jargon designed to confuse and embarrass people into going away and leaving him to his own devices. It is also foolish. For it has become increasingly apparent in the past few years that, unless all scientists do attempt to put their point of view, decisions which may adversely affect their own future – both as scientists and as citizens – will be taken on the basis of gut reaction rather than reasoned argument.

A decade and a half ago, in its first report, the American Association for the Advancement of Science's Committee on Science in the Promotion of Human Welfare concluded:

> . . . that the scientific community on its own initiative should assume an obligation to call to public attention those issues of public policy which relate to science, and to provide for the general public facts and estimates of the effects of alternative policies which the citizen must have if he is to participate intelligently in the solution of these problems. A citizenry thus informed is, we believe, the chief assurance that science will be devoted to the promotion of human welfare.[8]

## Assessing technology

Since these words were written there have been signs of change. Perhaps most significant is the way in which governments have realized that they must justify their policies in the light of informed criticism, for this has led to a rethinking of social goals at government level. In 1969 a committee convened by the U.S. National Academy of Sciences (NAS) published its report, *Technology: Processes of Assessment and Choice.*[9] This committee, which had been convened at the request of the subcommittee on Science, Research, and Development of the U.S. House of Representatives' Committee on Science and Astronautics, recommended establishment of new federal mechanisms to look broadly at the social consequences of technological actions.

Panel chairman Harvey Brooks, of Harvard University, and panelist Raymond Bowers described some of the panel's conclusions and recommendations. Looking at the U.S. pesticide experience, for example, the panel argued that:

8. 'Science and Human Welfare', *Science*, cxxxii, 1960, p. 68
9. Report of the National Academy of Sciences, Committee on Science and Astronautics, U.S. House of Representatives, U.S. Government Printing Office, July 1969.

> . . . although the pesticides have undoubtedly prevented a great many deaths from starvation and disease, it is now apparent that they have also inflicted unintended but widespread losses of fish and wildlife, and it is increasingly suspected that they are causing injury to man. The experience suggests that carefully designed experiments in the early days might have influenced the technology of pesticides before the nation was so committed to certain forms of pest control as to make any significant alteration of the technology extremely difficult.[10]

In most countries, no official body has until recently looked at the growth in numbers of automobiles. The domination of land transportation systems by the private car is a prime example of the dangers of the refusal to plan. It would be extremely difficult to make any significant alteration in transportation technology now, although the social costs of the private car become clearer (smoggier?) every day.

This type of consideration led the NAS panel to lay down an important criterion for assessing technology. 'The reversibility of an action should thus be counted as a major benefit, its irreversibility as a major cost.' But it appreciated that 'the achievement of a better system for assessing technology faces major obstacles. The society is ill equipped to handle conflicting interests. It does not know how to value in a quantitative way such goals as a clean environment and the preservation of future choices. Analytical tools are primitive and crucial knowledge is often missing.'[11]

The reorientation of science may encourage scientists to fill some of those crucial gaps; certainly, technology assessment is not concerned solely with forbidding activities:

> An effective system of assessing technology would as often stimulate the development and application of desirable new technologies and underemployed ones as it would give warning of possibly harmful side-effects. Many of the problems that are identified as undesirable results of technological development can also be seen as the result of failures to develop or apply technologies that would have mitigated the undesired effects. . . .
>
> The future of technology holds great promise for mankind if greater thought and effort are devoted to its development. If society persists in its present course, however, the future holds great peril, whether from the uncontrolled effects of technology itself or from an unreasoned political reaction against technological innovation – a

10. Harvey Brooks and Raymond Bowers, 'The Assessment of Technology,' *Scientific American*, ccxxii, 2, 1970, p. 15
11. Ibid, p. 18.

> reaction that could condemn mankind to poverty, frustration and the loss of freedom.[12]

A more recent report, which takes the problem of scientific and technological development even further, was published during the summer of 1971 by the Organization for Economic Co-operation and Development (OECD), of which the U.S., Japan and many European countries are members. This report, prepared by a special committee also chaired by Harvey Brooks, questioned many of the economic assumptions underlying the desirability of progress, and showed how the quality of life does not necessarily improve as incomes increase:

> We must . . . recognize that increasingly man cannot live by bread alone, and that the use of the economic system to contribute to human happiness rather than merely material satisfaction presents a challenge to the imagination that the developed countries must begin to take up seriously during the coming decade. . . . Science and technology are an integral part of social and economic development, and we believe that this implies a much closer relationship between policies for science and technology and all socio-economic concerns and government responsibilities than has existed in the past.[13]

The OECD committee took a gloomy view of the failure of governments to come to terms with the new requirements to be made of science and technology.

> The world-wide culture of educated youth, which is deeply concerned with ecological perspectives and is increasingly anti-materialistic, egalitarian, anti-meritocratic, and anti-bureacratic, could conceivably even adopt anti-rational views and could become much more influential in the next decade than our extrapolations suggest; . . . [the result of this might be] the retardation of scientific progress to the point where the world lacked the intellectual tools to cope with the complexity it has created.[14]

The danger of being unable to cope is not one which we should set aside lightly. In their report to the Club of Rome, summarizing the results of a systems analysis of possible world futures, Mesarovic and Pestel claimed that 'there is ample reason to believe that the problems of our time will *not* be solved in the routine course of events. For one thing, the numerous

12. Ibid, p. 20
13. *Science, Growth and Society*, Organization for Economic Co-operation and Development, Paris, 1971, pp. 28, 96.
14. Ibid, p. 24

crises of the present exist simultaneously and with a strongly woven interrelationship between them. We do not have the luxury of dealing with one crisis at a time. Furthermore, the scale and global character of the present crises differ from the nature and scale of most past crises.'[15]

Mesarovic and Pestel have also argued convincingly the need for multidisciplinary approaches to global crises:

> Traditionally, in order to understand what appeared to him to be diverse aspects of reality, man has developed different scientific disciplines: physics, chemistry, biology, technology, economics, the social and political sciences, philosophy, ethics, theology, etc. And in solving different problems man has relied on experience and expertise in relevant disciplines. But today's problems require knowledge from a number of, if not all, disciplines. For example, the solution of the world food supply crisis is a question not just of agronomy and economics, but of ecology, the physical and social sciences, and many others. How to increase the fertility of the soil, and the acreage of arable land, the question of landownerships, the organization of agriculture, etc, all are now critical and interdependent issues. Furthermore the solution depends on the population growth, since the problem is not in producing foodstuffs as such, but in producing food in the amount required by the existing population. Eventually, the availability of food in one part of the world and the desperate need for food in another will create a new international political situation; the very daily existence of the people in the needy parts of the world will depend on the decisions in other parts and over a long period of time, perhaps indefinitely.[16]

Regrettably, there are signs that, instead of taking a 'holistic' view of serious world problems, some vociferous groups of scientists and 'activists' are diverting public concern with scare stories, many of which prove later to have had little substance. Where this is the case, it is usually because the scientists involved opted for publicity before their data had been digested properly.

Many of the scare stories relate to the possible dangers from exposure to particular, usually industrially-produced, chemicals. Yet rarely is the danger as great as that to which many people expose themselves daily on a voluntary basis – through cigarette smoking, for example.

Professor Richard Scorer, a British meteorologist, has been particularly critical of what he describes as the new scientific ethic 'that we

15. Mesarovic and Pestel, *Mankind at the Turning Point*, pp. 10–11
16. Ibid, pp. 20–1.

should not take a risk if harmful effects are predicted scientifically unless we can disprove the theory in question'.[17]

A recent example of what Scorer calls 'the environmental jitters' is the ozone-chlorofluorocarbon controversy. In 1971, using an extremely sensitive analytical technique which he had invented, the British chemist Jim Lovelock detected traces of chlorofluorocarbons in the atmosphere. These compounds do not occur naturally, but are manufactured for use as refrigerants in refrigerators and as propellants in many aerosol spray cans. Lovelock looked for them in the atmosphere because he thought they might provide useful 'markers' for the study of air movements.

Chlorofluorocarbons are stable, relatively inert chemicals, which is why they are successful as aerosol propellants. However, when exposed to ultraviolent radiation, they can break down to produce chlorine atoms. These atoms react rapidly, in a laboratory, with ozone, breaking it down to ordinary bimolecular oxygen, and regenerating themselves in the process.

In the upper atmosphere is a region called the 'ozone layer' and, during 1974, a group of American chemists suggested that chlorofluorocarbons could seriously deplete this layer and consequently increase the penetration of ultraviolet radiation from the Sun to ground level. One suggested result would be an increase in the incidence of skin cancers among white people; other possible effects include damage to agricultural crops and livestock. Gradually, public reaction was stimulated by continued warnings from the scientists involved and from a large number of vociferous non-scientists who were 'concerned' about what the scientists had told them, but apparently not concerned enough to listen to the serious objections raised to the theory by other scientists. In June 1975, a U.S. government taskforce recommended that, unless new evidence was produced against the theory, aerosols using chlorofluorocarbons should be banned. Early in 1978, bans on non-essential aerosol uses and bulk manufacture of these materials from late 1978 were announced by U.S. government agencies.

On one level, the action taken by politicians may seem highly responsible. They listened to scientists and acted on evidence supplied to them. At the same time, however, they appear to have ignored a substantial amount of contradictory evidence. For example, the 'ozone layer' is actually a highly complex and variable region in the upper atmosphere. By and large, it increases in concentration with increasing latitude, but it is by no means uniform and nobody knows by how much its average concentration fluctuates naturally with time, nor what the long-term

17. Richard Scorer, 'The danger of environmental jitters', *New Scientist*, lxvi, 1975, p. 702

effects of such fluctuations are. 'The most important criticism of the whole idea' of the layer's destruction by man-made chemicals, wrote Scorer, 'is that it requires the stratosphere to be constituted more simply than it could possibly be'.[18] Further, the hypothesis that extra uv would cause an increase in skin cancer is based on weak statistical evidence.

Another factor ignored by the theorists is the effect of the very large amounts of natural chlorocarbons released into the atmosphere each year. No one, apparently, had bothered to find out whether there were many naturally-occurring atmospheric chemicals which might generate chlorine atoms until after the scare about chlorofluorocarbons had gathered momentum. Yet, once looked for, they were found to be abundant and to occur in quantities vastly greater than the industrially produced materials. Jim Lovelock, who also discovered the presence of the natural chlorocarbons in the atmosphere (and testified as much to a Congressional committee two months before the U.S. taskforce produced its report), has even suggested that their production might be part of a cycle which functions naturally to keep ozone concentration down, and that we perhaps ought to look into the possible dangers of higher ozone levels, such as increased brittleness of bones. The chemistry of the atmosphere remains too complex to provide clear-cut answers.

It is difficult to believe that the time and effort which went into publicizing the ozone scare could not have been spent better on more serious matters. As Scorer argued:

> There will very probably be some interesting new facts elucidated about chlorine in the stratosphere in the next few years, but more important than knowing them is to have a confident philosophy about the environment which will stop important decisions being made on the basis of the jitters, and which will get people thinking about the important aspects of the present human predicament and how we are going to face them during the next half century.[19]

Professor Abdus Salam, director of the International Centre for Theoretical Physics in Trieste, has said that 'by and large, we do possess the information necessary to solve the problems of hunger and most disease. We simply do not take the right political, economic and social decisions to put the requisite policies into effect and give the requisite technologies the scope they should have.'[20]

18. Ibid, p. 703.
19. Ibid.
20. 'Organized research and common sense', an interview with Abdus Salam, *Impact of Science on Society*, xxv, 1975, p.55.

During the recent past, many scientists have been surprised and hurt by the widespread criticism that their activities have aroused. It is easy to see how, in the face of such criticism, some scientists may have been tempted to restore their esteem by exposing scientific 'dangers' to public scrutiny, and thus hoping to convince the public that they are good guys. Nevertheless, such an attitude reflects a dubious morality in so far as it exposes the scientific enterprise to ridicule when their prematurely publicized theories are shown to be half-baked, and even more in so far as it diverts public attention from the fundamental decisions that must be taken if we are to 'give the requisite technologies the scope they should have'. For example, it would have been a far more valuable enterprise to make people aware of the general undesirability of aerosol-packaged products in a world that has only limited material and intellectual resources – for the production of aerosol cans and the formulation of new products consumes both to an extent out of all proportion to the value of the end product in terms of its fulfilment of human needs – than to single out one part of the package for possibly spurious denigration.

## Forecasts

The preceding paragraphs are intended to show not only the importance of growing public awareness of – and, to some extent, disillusionment with – scientific advance, but also the potential which a 'holistic' approach, involving all disciplines, can have for mankind's future. From such an approach, however, springs the question: has chemistry a future? Growing demands made upon scientists to show both responsibility and responsiveness will affect chemists as much as any others. One result may be to hasten the withering away of chemistry as an active science clearly different from biology or physics. In the preceding pages it has been impossible at times not to step outside the hazy borders of chemistry; without doing so, no sensible picture would have emerged. Thirty years from now those borders may be even less clearly discernible.

There is a second reason for stressing the mood of our age and its effect on science, rather than making concrete predictions. A scientific discovery is basically unpredictable. It is a venture into the unknown. This, you may say, only affects fundamental research; surely we can predict how industrial chemistry will develop? Even here the answer is only a partial yes. Had there been 'futurologists' in the nineteenth century, predicting on the basis of current trends, at least one of them would surely have pointed with horror to the rising demand for horses as a means of transportation, and claimed that by 1950 there would be a shortage of horses so severe that civilization would grind to a halt. The invention of the automobile would have rendered any such prediction incredibly silly, just as, in our lifetimes, a sudden breakthrough in the development of controlled nuclear fusion as a source of energy could make all present predictions about energy supply look ridiculous.

Those who are not swayed by hypothetical examples may be interested in the following quotations from real people, just a few of those used to garnish a 1969 article on technological forecasting in the American Chemical Society's magazine, *Chemical & Engineering News*:

> This is the biggest fool thing we have ever done. The [atomic] bomb will never go off, and I speak as an expert in explosives—Admiral William Leahy, 1945.

> You could put in this room [his office], de Forest [inventor of the vacuum tube] and all the radiotelephone apparatus that the country will ever need—W. W. Dean, President of Dean Telephone Company, 1907.

> I have always opposed consistently high-tension and alternating systems of electric lighting . . . not only on account of danger but because of their general unreliability and unsuitability for any general system of distribution—Thomas A. Edison, 1889.[21]

These three examples from the past show the danger of saying what science and technology will never be able to do. But even the optimistic can be made to look foolish. Confident predictions of future developments are frequently outstripped by reality, yet people still make predictions.

In the following paragraphs I hope to identify some trends, possibilities and likelihoods. Since the best way to avoid being proved wrong at a later date seems to be to keep this identification as general as possible, the rest of this book can in no way be called a recipe for the future; at most it is an indication of flavour.

Underlying chemistry are a number of theories, the importance of which to the development of the subject cannot be overestimated. However, in recent years chemistry has tended to race ahead of theory in some areas, notably the synthesis of complex organic molecules. There are signs of a change here. Several new theoretical concepts of the last few years are simplifying the approach to synthetic chemistry. The Woodward-Hoffmann rules, worked out by American chemists Robert Woodward and Roald Hoffmann, predict the stereochemistry of the products to be

21. David M. Kiefer, 'The Futures Business,' *Chemical & Engineering News*, xlvii, 33, 1969, pp. 62–75.

expected from certain reactions on the basis of orbital interactions. Similar orbital considerations occur in the concept of antiaromaticity. In systems of alternating single and double bonds, electrons are delocalized (e.g. benzene) and the compounds stabilized; antiaromatic systems are, in a sense, the reverse: the number of electrons involved in bonds has a destabilizing effect, which alters their reactivity dramatically. Despite these and other theories, such as the concept of 'soft' and 'hard' acids and bases, which covers both organic and inorganic molecules, there is a need for more theoretical rationalization of the ever-increasing mass of chemical data. Not only could new and more general theories, for example to explain the phenomenon of catalysis, have beneficial economic effects, the development of 'tidying up' theories would help to restore a unity that is presently lacking from our consideration of molecules.

Ultimately, one can hope for a theory which will permit working out the physical and chemical properties of 'imaginary' compounds on the basis of molecular orbital theory. Having worked out the properties, one could also hope to work out, in fine detail, how to make a compound, if such a task seemed useful, and what the reaction rates and yields of the compound and the various intermediates would be without ever having to dirty one's hands at the laboratory bench.

Michael Dewar and his colleagues at the University of Texas, Austin, have produced some interesting results in this field with a programme called MINDO/3. The acronym stands for Modified Intermediate Neglect of Differential Overlap. Based on quantum theory, with some necessary simplifications, and some basic chemical intuition, MINDO/3 calculates many of the properties of relatively simple molecules rather well. The system is still being developed but it has 'already made some interesting contributions to the theory of chemiluminiscent processes' and provides 'for the first time an apparently reliable and thoroughly tested procedure that can be applied at reasonable cost to real chemical problems'.[22] The phrase 'reasonable cost' is important, for computer-simulated chemistry will never be much used if it costs more to do than the actual experiments.

In experimental chemistry, can one expect new instruments or separation techniques that will advance the field as much as, say, nuclear magnetic resonance spectroscopy and gas chromatography have done? Advances in instrumentation occur regularly, most of them small, but some big. It is unlikely that anyone writing a book such as this 20 years ago would have predicted the way in which electron spin resonance spectroscopy would grow into an important chemical tool. Equally unexpected

22. Michael J. S. Dewar, 'Quantum organic chemistry', *Science*, lxxxvii, 1975, p. 1044.

would have been the effects that the then new invention, the laser, would have on chemistry, both in diversifying spectroscopic techniques and in the study of very fast reactions. Can we even now predict the advances that might be made as a result of holographic microscopy? In September 1974, Lawrence Bartell and Charles Ritz of the University of Michigan described a holographic microscope with a theoretical resolving power of 0·08Å (about one-tenth the separation of the two nuclei in a hydrogen atom) and published photographs of the electron density in neon and argon atoms at magnifications of 260,000,000.

Other new techniques in recent years have shown the possibility of directly probing the stereochemistry of large molecules spectroscopically, and quantitatively analysing very complex molecules on the basis of a few millionths of a gramme of material. Going on past experience, the best bet for those inclined to gamble on new instrumentation and techniques is to find out what new phenomena the physicists are studying and see if there are any that chemists might put to use. In the light of this, it would be foolish to say that no revolutionary separational method will be discovered in the next few years (laser separation of isotopes is one major possibility); but more likely is the further improvement and sophistication of existing techniques. For example, affinity chromatography – a form of chromatography for separating biologically active molecules which uses the specific reactivities of these molecules as a means of separation – is already making enormous strides. As well as being a boon to the chemist who wishes to isolate, for example, a particular enzyme from a group of enzymes, this technique could be valuable in medicine, providing means of separating materials such as steroids from mixtures on an industrial scale. Meanwhile, electron capture detector gas chromatography, the technique pioneered by Jim Lovelock (see p. 217), is still being developed – and may soon reach the stage where it can detect and identify molecules present in a gas at concentrations as low as one part in a thousand billion ($10^{15}$).

The marriage of chemistry and biology in recent years has moved the pharmaceutical industry away from the concept of synthetic drugs, which have little relationship to any biological molecule, toward the use of natural materials, or slight modifications of them. However, as more is learned by biologists about the shape and structure of receptor sites – the target molecules with which drugs, hormones, poisons, and so on, react to produce their characteristic actions – it should become increasingly possible to design molecules with specific therapeutic effects, and thus put drug research on a more rational base than it has occupied in the past.

Among the biological substances which, according to some predictions, are likely to be available before the end of the century are drugs to increase intelligence and to treat mental illness. The study of brain chemistry is still in its infancy, and it is not yet possible to draw guidelines for future activities in this field. For the last few years there has been controversy about whether learning can be transferred from one animal to another through injections of brain chemicals, but no conclusive evidence has yet been produced for or against. Another possibility that has been suggested is chemical stimulation of the body to produce replacement limbs and organs. Every cell in the human body contains in its gene complement all the instructions required to produce that individual. As already mentioned (p. 120), one of the key questions of molecular biology at the present is: Why and how is most of the gene content of each cell switched off? Subsidiary questions are: What is the mechanism by which a particular part of a gene is switched off at a particular time? Is this process reversible? If the cells in the stump of an amputated limb could be triggered to begin dividing and differentiating in the way they did when the limb first developed, a wide range of repairs to bodies damaged in accidents would become possible.

At the extreme of this technique is the possibility of genetic engineering – changing the molecular structure of the gene so that defects are ironed out, making it possible, for example, to cure the inborn errors of metabolism that produce sickle-cell anaemia and phenylketonuria. Much of the technique of genetic engineering will be based in biology, but it seems likely that chemistry will play a part, or at least be affected by the results. It has been suggested that, in the future, agriculturists might engineer plants which can fix their own nitrogen and are pest and disease resistant, thus requiring less fertilizer and no pesticide. Of course genetic engineering, as a reality, is a double-edged sword. It need not be used for the welfare of individuals, and could be used in a variety of unpleasant ways. Consequently, it is one area of scientific possibility in which the general public must attempt to keep up with and understand the latest developments.

More significant at the moment is the possible danger from uncontrolled experimentation in this field. Scientists involved are concerned not so much with the ultimate social uses to which genetic engineering might be applied, as with the possibility of accidentally engineering some hazardous microorganism, which might then escape and wreak havoc in unprotected populations of animals, plants or humans. A number of countries have adopted, or are formulating, codes of practice for the conduct of research in this highly-sensitive field.

A less controversial area is materials science, where it seems likely that the development of new materials for a variety of purposes will continue along the molecular engineering lines described in Chapter 6. There is growing interest in the field of inorganic polymers, which could lead to widespread commercial applications within the next few years. Sulphur-nitrogen polymers, for example, are being investigated because of their electrical and optical properties, which mimic those of metals; phosphorus-nitrogen polymers, on the other hand, could provide new electrical insulators, and flame retardant materials to replace some of the organic polymers, such as polyurethane foams, which are possible fire hazards when used in domestic furnishings.

Another part of materials science in which there could be remarkable developments during the next few years is bioengineering – the design and manufacture of spare parts for medicine and surgery. This is not a new field (see p. 171), but it is one in which industry is taking considerable interest. ICI, for example, has established a scientific group in conjunction with doctors from Liverpool University Medical School to explore various possibilities. There is scope in the field for new developments in polymer chemistry that would lead to applications varying from the mundane – such as better materials for filling dental cavities – to the almost fantastical. In the latter category one might include temporary structures which hold tissues in place while they heal and which then gradually dissolve away, and the possibility of polymer matrices which could be impregnated with drugs, so that a person in need of continuous medication could have a pharmaceutical pellet inserted surgically once a year, instead of having to take tablets or injections daily. The principle of a matrix that would slowly release a biologically active agent could be applied in other ways – as a once-yearly replacement for the contraceptive pill, for example.

Back from the frontiers of materials science, the balance of materials used in different industrial processes will alter in accord with various socioeconomic factors. Cellulose in its various forms – as wood, or as discarded wrappings, for example – could become an important raw material for the chemical industry. New techniques for the extraction of metals may be needed as we are forced to use lower-grade ores to keep up with the demand for these resources – here the chemist, or the microbiologist, may have an important role to play.

Various other predictions have been made by different groups of experts. According to some of these, there is a strong likelihood that it will be possible before the end of the century to manufacture elements from subatomic particles on a commercial basis, and possible not only to make living – or at least self-replicating – chemical systems, but also to manufacture protein synthetically. Any of these prospects might have immense social effects, but I still believe that the most exciting discoveries of the next thirty years have not yet been predicted.

How many people, thirty years ago, would have predicted the invention and discovery of noble gas compounds, the transistor, the laser, holography, and gene synthesis? A few may have had inklings of one or two of these, but no more. Yet they are all now with us, and all have sprung from the framework of modern science. Although there is much we cannot say about the chemistry of tomorrow, we can at least be sure that most of it will lie within the broad conceptual framework which this book has attempted to describe.

# Suggestions for further reading

## Chapter 1

Cottrell, T. L., *Chemistry,* 2nd ed., Oxford University Press, New York 1970

Jones, G. O., J. Rotblat and G. J. Whitrow, *Atoms and the Universe,* 3rd revised ed., Penguin, Harmondsworth, 1973

Pimentel, George C. and Richard D. Spratley, *Chemical Bonding Clarified through Quantum Mechanics,* Holden-Day, San Francisco, 1969

Selig, Henry, John G. Malm and Howard H. Claassen, 'Chemistry of the noble gases', *Scientific American,* ccx, 5, pp. 66–76

Taylor, John, *New Worlds in Physics,* Faber, London, 1974

## Chapter 2

Allen, G., 'Modern methods for the determination of molecular structure', J. G. Stark (ed.), *Modern Chemistry,* Penguin, Baltimore, 1970, pp. 56–75

Angrist, Stanley W., and Loren G. Hepler, *Order and Chaos: Laws of Energy and Entropy,* Basic Books, New York, 1967

Betteridge, D., and H. E. Hallam, *Modern Analytical Methods,* Chemical Society, London, 1972

Campbell, J. A., 'Why chemical reactions occur', J. G. Stark (ed.), *Modern Chemistry,* Penguin, Baltimore, 1970, pp. 113–1140

Chalmers, R. A., *Aspects of Analytical Chemistry,* Oliver & Boyd, Edinburgh and London, 1968

McLauchlan, Keith, and Roger Sealy, 'Photochemistry in a spin', *New Scientist,* lxix, 1976, p. 433

Pimentel, George C., and Richard D. Spratley, *Understanding Chemical Thermodynamics,* Holden-Day, San Francisco, 1969

**Chapter 3**

Ahrens, L. H., 'The chemical bond and the geochemical distribution of the elements', *Chemistry in Britain,* II, 1966, pp. 14–19
Barrer, R. M., 'Mineral synthesis by the hydrothermal technique', *Chemistry in Britain,* II, 1966, p. 380.
Bernal, J. D., *The Origin of Life,* Universe Books, New York, 1967
Calvin, Melvin, *Chemical Evolution,* Oxford University Press, New York, 1969
Eigen, M. and R. Winkler-Oswatitsch, 'The game of evolution', *Interdisciplinary Science Reviews,* i, 1976, p. 19
Eshleman, Von R., 'The atmospheres of Mars and Venus', *Scientific American,* ccxx, 3, 1969, pp. 78–88
Fox, Sidney W., Kaoru Harada, Gottfried Krampits and George Mueller, 'Chemical origins of life', *Chemical Engineering and Chemical News,* 25, 1970, pp. 80–94
Fyfe, W. F., *Geochemistry,* Clarendon Press, Oxford, 1974
Keosian, John S., *The Origin of Life,* New York, Reinhold, 1965
Lewis, John S., 'The origin of the planets and the satellites', *Technology Review,* lxxvi, 1, 1973, p. 20
———, 'The chemistry of the solar system', *Scientific American* ccxxx, 4, 1974, p. 50
Marquand, J. *Life: its Nature, Origins and Distribution,* W. W. Norton, New York, 1968
National Academy of Sciences, *Orientations in Geochemistry,* National Academy of Sciences, Washington, D. C., 1973
'The Solar System', *Scientific American,* ccxxxiii, 3, 1975 (a single-topic issue)
Stubbs, Peter (ed.), *New Science in the Solar System,* IPC, London, 1975
Ulbricht, Tilo, 'Left and right are not equal', *New Scientist,* lxvii, 1975, p. 479
Wickramasinghe, Chandra, 'Where life begins?' *New Scientist,* lxxiv, 1977, p. 119
Wild, J. P. (ed.), *In the Beginning . . . The Origins of Planets and Life,* Australian Academy of Science, Canberra, 1974. This contains articles by Prentice, Ringwood and Orgel among others.

**Chapter 4**

Brown, E. G. *An Introduction of Biochemistry,* Royal Institute of Chemistry, London, 1971
Butler, J. A. V., *The Life Process,* Basic Books, New York, 1970
Calvin, Melvin, and Margaret J. Jorgenson (eds.), *Bio-organic Chemistry,* Freeman, San Francisco,1969. This collection of readings from *Scientific American* includes articles on insulin, hemoglobin, DNA structure, sequencing of alanyl t-RNA, pheromones, insect attractants, stereo chemical theory of odour, analgesic drugs, photosynthesis and molecular isomers in vision.
Dayhoff, Margaret O. 'Computer analysis of protein evolution', *Scientific American,* ccxxi, 1, 1969, pp. 86–95
Merrifield, R. B., 'The automatic synthesis of protein', *Scientific American,* ccxviii, 3, 1968, pp. 56–74

O'Malley, Bert W., and William T. Schrader, 'The receptors of steroid hormones', *Scientific American,* ccxxxiv, 2, 1976, p. 32
Perutz, Max, 'Hemoglobin—the molecular lung', *New Scientist,* vol. 50, 1971, pp. 676–679
———, 'Hemoglobin—genetic abnormalities', *New Scientist,* vol. 50, 1971, pp. 762–765
Sullivan, Navin, *The Message of the Genes,* Basic Books, New York, 1967

### Chapter 5

Jones, D. G. (ed.), *Chemistry and Industry,* Oxford University Press, New York, 1967
Lushington, Roger, *Plastics and You,* Pan Books, London, 1967
Sherwood, Martin, 'Chemicals and the oil crisis', *New Scientist,* lx, 1973
Waddams, A. L. *Chemicals from Petroleum,* 3rd ed., John Murray, London, 1973
Wynne, M. D., *Chemical Processing in Industry,* Royal Institute of Chemistry, London, 1970

### Chapter 6

Chedd, Graham, *Half-way Elements: the Technology of the Metalloids,* Doubleday, New York, 1968
Fishlock, David, The New Materials, Basic Books, New York, 1967
Fleming, Stuart, 'Science detects the forgeries', *New Scientist,* lxviii, 1975, p. 567
Gregory, J. G. (ed.), *Modern Chemistry in Industry,* Society of Chemical Industry, London, 1968
Imrie, Frazer, 'Single-cell protein from agricultural wastes', *New Scientist,* lxvi, p. 458
Libby, W. F., 'Radiocarbon dating', *Chemistry in Britain,* v. 1969, pp. 548–552
Pyke, Magnus, *Synthetic Food,* St. Martin's Press, New York, 1970
Rowe, Geoffrey W., 'The chemistry of tribology (friction, lubrication and wear)', *Royal Institute of Chemistry Reviews,* I, 1968, pp. 135–204
Sherwood, Martin, 'Single-cell protein comes of age', *New Scientist,* lxiv, p. 634
Werner, A. E. Scientific methods in art and archaeology', *Chemistry in Britain,* vi, 1970, pp. 55–59
Youle, P. V., and J. R. Stammers, 'Polymers in an age of sense', *Chemistry in Britain,* xii, p. 53

### Chapter 7

Brooks, Harvey and Raymond Bowers, 'The assessment of technology', *Scientific American,* ccxxii, 2, 1970, pp. 13–22
Commoner, Barry, *Science and Survival,* Ballantine, New York, 1971
'Forecasting the future', *Science Journal,* III, 10, 1967 (a single-topic issue)

Hamblin, Lynette, *Pollution, and the World Crisis,* Barnes and Noble, New York, 1971
Kenward, Michael, *Potential Energy,* Cambridge University Press, Cambridge, 1976
Kuhn, Thomas S., *The Structure of the Scientific Revolution,* 2nd ed., University of Chicago Press, Chicago, 1970
Mesarovic, Mihajlo, and Eduard Pestel, *Mankind at the Turning Point* (The Second Report of the Club of Rome), Dutton, New York, 1974
Naudet, Roger, 'The Oklo nuclear reactors: 1800 million years ago', *Interdisciplinary Science Review,* i, 1976, p. 72
Patterson, Walter C., *Nuclear Power,* Penguin, Harmondsworth, 1976
Ravetz, J. R., *Scientific Knowledge and its Social Problems,* Oxford University Press, New York, 1971
'Science and common sense', *Impact of Science on Society,* xxv, 1, 1975 (25th Anniversary issue)
*Science, Growth and Society,* Organization for Economic Cooperation and Development, Paris
Sherwood, Martin, 'Single-cell protein comes of age', *New Scientist,* lii, 1971, pp. 102–103
Woodwell, George M., The carbon dioxide question', *Scientific American,* ccxxxviii, 1, p. 34.

## Index